)0

Residential Electrical Estimating

By John E. Traister

Craftsman Book Company
6058 Corte del Cedro
Carlsbad, CA 92009

Library of Congress Cataloging-in-Publication Data

Traister, John E.
Residential electrical estimating / by John E. Traister.
p. cm.
Includes index.
ISBN 1-57218-013-7
1. Electrical engineering--Estimates. 2. Electrical wiring, Interior--
-Estimates. I. Title.
TK435.T73 1995
621.319'24'0299--dc20 95-9803
CIP

Contents

Preface

Some years ago, when I was an electrical contractor bidding mostly residential and small commercial installations, several estimating methods were tried:

Quantity survey — Quantities of all materials were first listed for a particular project. Then labor units were added to each material item. The material was priced, and a hourly dollar rate given to the total worker-hours. A "selling price" for the given project was determined after adding overhead and profit. This method provided very accurate bids, but it was very time-consuming. Oftentimes, my workers could have a house roughed-in in the same amount of time it took to compile an accurate estimate.

Another problem with the quantity-survey method was losing on-the-spot work; that is, a customer may ask for an estimate for some extra work or a couple of additional circuits to an existing installation. By the time I made a detailed estimate, the customer frequently had another contractor already doing the work.

Square-foot method — I used three different levels to price residential work by this method:

- Good
- Better
- Best

The *Good* projects were dwellings that were wired to minimum standards as set forth in the *National Electrical Code®* (*NEC*). The per-square-foot price was arrived at by calculating the total installation costs of several projects in this category, and then dividing by the area of the house. In all cases, costs of lighting fixtures, chimes, electric heat and similar items were added to the total price.

The *Better* projects were similar to the *Good*, except for a few more outlets than were required by the *NEC*, and perhaps rigid conduit was used for the service mast instead of Type S.E. cable.

Houses falling into the *Best* category utilized several more outlets than were required by the *NEC*, usually had higher-grade wiring devices, and a lot of electrical appliances to connect. Again, the cost of lighting fixtures, electric heat and the like were added as extras.

While this method saved a lot of estimating time, departure from the actual cost of some projects was sometimes great. Too much of a variation for the highly-competitive business of residential electrical contracting. There was also the problem of calculating the cost of extra work, such as a couple of additional wall receptacles, or the connection of an electric water heater. When it came to renovation projects, or work on existing structures, the square-foot method was of no use at all; I had to revert back to the quantity-survey method.

I finally arrived at a unit-price method that gave highly accurate estimates, was extremely fast, and could also be used by workers to calculate the cost of extra work without an estimator being present.

In general, the unit-pricing method combines amounts and types of material used for one unit; that is, a conventional wall receptacle, a small-appliance receptacle, a range receptacle, etc. Each group or unit is assigned a material and labor price. The number of units are then categorized, and the total number multiplied by the group price for a total cost of the installation. Overhead, profit, and direct costs are added to arrive at a total "selling" price for the project.

This book describes the unit-pricing method in detail, including estimating forms that may be used to estimate residential electrical work. The tables in this book are further designed to provide a basis for each contractor to develop his or her own unit costs and selling prices. All items have been job-tested and used to estimate at least 1000 residential electrical installations, ranging in price from small vacation homes to more elaborate homes costing over $300,000.

Chapter 1

Introduction to Electrical Estimating

Sound cost estimating of any electrical installation usually consists of a complete takeoff (or quantity survey) of all materials and equipment required for a complete installation, and then the calculation of the total labor required to install equipment and materials. To the cost of materials, equipment, and labor are added all direct expenses, variable job factors, taxes, overhead, and profit to determine a "selling price" for the project. If these procedures are intelligently performed and combined with good job management, the estimate should compare very favorably with the actual installation cost.

The steps necessary to prepare a cost estimate for a given electrical installation traditionally run as follows:

Takeoff: The count of all lighting fixtures, outlets, and similar items, and the measurement of all branch-circuit wiring, feeders, service raceways, etc.

Listing the Material: All items accounted for in the takeoff should be listed in an orderly sequence on a standard pricing sheet.

Applying Labor Units: Determining the proper labor unit from proven labor-unit tables and applying them to the various materials or labor operation under the labor-unit column on the pricing sheet.

Finalizing: The summation of material dollars and labor hours, the assignment of job factors and dollar values to labor hours, and the determination of overhead and profit.

Unit Pricing

Some electrical systems — like residential and some small commercial projects — lend themselves to standard materials and operations, making unit pricing practical. This method seems to be the simplest, fastest, and most accurate way to compile residential estimates within a reasonable time frame. The unit-pricing method combines into groups the electrical materials that are commonly used together. Each group is assigned a material and labor price (prime cost) to be used during the estimating process. Overhead, profit and direct job expense are then added to the previous total for a final selling price for the given project. Let's look at an example.

To find the price of a conventional wall-mounted duplex receptacle in a house containing, say, 50 such receptacles, we completed the installation and then made a one-on-one count of all materials used. Furthermore, we kept accurate records of worker-hours required to install only these 50 receptacles. The results per receptacle were as follows:

- One outlet box
- Two 16-L nails to secure box to wall studs
- One grounding clip
- One 15-ampere duplex receptacle (residential grade)
- One receptacle cover plate
- 23 feet of 14-2 w/ground Type NM cable (Romex)

During the tests, the 50 receptacles required 1150 feet of Romex cable which included runs from outlet to outlet, and all homeruns to the main panelboard. Some cable runs were only a few feet, while others were perhaps 30 feet or more. To find the average per outlet, divide the total footage of cable by the number of outlets.

$$\frac{1150}{50} = 23 \text{ feet}$$

Thus, it was determined that each duplex receptacle outlet required 23 feet (average) of Type NM cable. Obviously, some houses will require a little more, while others will require less. However, over the span of a few dozen houses, the difference in the amount of cable used is usually not worth mentioning, unless several unusual situations develop.

The cost of material for one outlet was then calculated, using current prices from manufacturers or electrical suppliers. Let's say the material for one standard duplex wall receptacle was $10. Consequently, this is the prime cost of materials for one standard duplex receptacle.

Once the material unit was priced, the next step was to calculate the average labor-unit price which entails determining the time required to install each outlet, and then multiplying the time by the worker's rate per hour.

During our tests, we discovered that it took 40.5 worker-hours to install the 50 wall-mounted duplex receptacles in new construction. This figure may seem high to electricians experienced in residential construction. However, let's break this figure down to see where the time was spent.

Unloading materials and tools from truck, setting up extension cords, and drinking cup of coffee	1 Worker-Hour
Layout and installing boxes	4.5 Worker-Hours

Drilling holes through sole plate under each appropriate outlet box	1 Worker-Hour
Drilling holes through joists and studs	2 Worker-Hours
Pulling cable, securing cable with staples, stripping ends of cables, securing with staples, stripping ends of wires, securing under cable clamps in outlet boxes	10 Worker-Hours
Making splices in outlet boxes, securing ground wires under clips, and placing wires neatly inside of outlet boxes	5 Worker-Hours
Packing up tools and materials; loading truck	1 Worker-Hour
Total Worker-Hours for rough-in	24.5

Trim-Out Stage

Installing duplex receptacles and plates, testing system for ground-faults	16 Worker-Hours
Total Worker-Hours for project	**40.5 Worker-Hours**

To find the average time required for each receptacle, divide the total worker-hours by the number of receptacles.

$$\frac{40.5}{50} = 0.81 \text{ worker-hours per receptacle}$$

This time per unit is now converted to a cost per unit. To make this conversion, the cost per labor hour for the electricians doing the work must be determined. This rate, of course, will vary in different parts of the country, and also from contractor to contractor. First determine the base wage; that is, what each electrician is paid per hour. Then calculate fringe benefits, employee insurance, association assessments, F.I.C.A. wages, and other applicable taxes that must be paid by the employer.

Again, the fringe benefits paid by employers and the required state and federal taxes will vary, but in most cases, the contractor can figure about 30% additional above the hourly wage. Consequently, if electricians are being paid, say, $15.00 per hour, multiply 1.30 x 15.00 = $19.50. These costs must not be overlooked.

Total Prime Cost

Now that the labor cost per hour has been determined ($19.50), the labor cost per unit may be determined. Multiply the hourly cost x the time required for each receptacle (.81 worker-hours).

$19.50 x .81 = $15.80 per receptacle

Since it has already been determined that the material cost per outlet is $10 and the labor cost per outlet is $15.80, the total prime cost for a conventional duplex wall receptacle is $25.80. That is, if all of the following factors are valid:

- No. 14-2 w/ground NM cable is used.
- The work is standard new work; not renovation or unusual situations.
- The total material cost is $10.00 per outlet.
- The total labor cost per hour is $19.50.

Varying any of the above factors will also vary the installation cost. For example, if duplex receptacles are to be used in small-appliance circuits, the wire size must be increased to No. 12 AWG. Consequently, a little more labor will be involved and also the cost of No. 12-2 w/ground NM cable will be higher than the No. 14 AWG. If the receptacles are of the ground-fault circuit-interrupter type, the cost of materials will increase, or if the receptacles require weatherproof covers for use outdoors, the cost of materials will also increase. Therefore, a separate table is needed for each type of outlet and also for each type of wiring method and working conditions. Tables for nearly every conceivable wiring application are included in this book. Furthermore, all tables in this book are specifically for residential installations — for both new and renovation work.

Wiring Methods

The experienced electrician or contractor readily recognizes, within certain limits, the type of wiring method that should be used. However, always check the local code requirements when selecting a wiring method. The *National Electrical Code® (NEC)* gives minimum requirements for the practical safeguarding of persons and property from hazards arising from the use of electricity. These minimum requirements are not necessarily efficient, convenient, or adequate for good service or future expansion of electrical use. Some local building codes require electrical installations that surpass the requirements of the *NEC*, so always check for any deviations from the *NEC* in the area in which you are working.

The tables in this book are grouped together according to wiring method and are of two types. One is for unit prime cost determination and is followed at the end of the chapter by the corresponding table for calculating unit selling prices.

Data are included for complete branch circuits, feeders, special-purpose circuits and electric services. The tables cover all the usual wiring methods that are normally found in residential occupancies; that is, armored cable (BX), nonmetallic-sheathed cable (Romex), electrical metallic tubing (EMT), rigid conduit, PVC conduit, and surface metal raceway.

It is possible that on some rare installations, circuits or installation situations not described will be encountered. In this event, it is recommended that the data for a comparable installation be used.

Because of the wide variation in size, type, quality and price of lighting fixtures, their costs and installation must be included in the estimate as separate items.

The data in the tables are based on average complete installations and apply only to complete average installations. The information should not be used without upward adjustment to determine cost or selling price for the installation of only one or a few outlets. All the elements of cost — material, labor, direct job expense and overhead — will differ for complete and single-outlet installations.

When a limited number of outlets are to be provided on a renovation job, the person preparing the estimate should either increase each unit price used or add a contingency factor to the total, based upon the regular calculated unit prices according to his or her judgment of the situation.

Increasing Business With Unit Selling Pricing

The experiences of residential electrical contractors have shown that it is often necessary to sell a customer on the first visit. Taking the time to prepare an estimate on the basis of making a detailed takeoff and material list gives the prospective customer the opportunity to change his or her mind or to give the job to a more aggressive contractor. Also, the relatively smaller total price per job for residential work, both for new work and updating existing installations, is not sufficient to allow for a detailed estimate.

When properly used, a set of unit prices for renovation work provides a good selling tool. When, in the course of making up an estimate for upgrading an existing installation in the presence of a homeowner, referring to a precalculated set of unit selling prices often impresses the homeowner, and often a contract can be signed on the spot.

Skill Of Workers

Electrical contractors specializing in house wiring have highly skilled technicians. The labor units in the estimating tables are based on very efficient electricians, familiar with residential wiring methods. Contractors not regularly employing such workers must make additional allowances accordingly.

Variable Labor Costs

Even though residential installations do approach a standard, there are variables — especially with regard to labor cost for renovation work that must be considered and for which adjustments must be made.

The principal causes of variation in labor cost between new house wiring and different types of renovation work are:

- The extent of concealed wiring required.
- The nature of the building construction and building materials encountered.
- The accessibility of the specified working area.
- The difficulty encountered by the workers due to the inaccessible and restricted work spaces such as attics, low basements or crawl spaces, etc.

The tables in this book are designed so that the variations in labor cost are considered when preparing unit selling prices.

Variations Of Installation Situations

There are an almost infinite number of installation situations that can affect the labor cost for any given type of outlet, circuit, feeder, etc. The total effect of these varying conditions adds up to a greater or lesser amount of time for the work required to install any given item. An entirely different set of variable conditions might affect different outlets in different dwellings or the same general type of outlet installed in different locations in the same dwelling and yet the total labor required per item might be the same.

Basis Of Installation Situation Groups

To allow for the variable conditions and at the same time keep the listing of data to a practical length, conditions have been grouped into four installation situation groups, with one labor unit for each group to cover all situations in the group. These groups are designated 1, 2, 3 and 4 with Group 1 representing the installations requiring the least expenditure of worker-hours and Group 4 requiring the most.

Let's look at how installation variables can affect labor costs. Assume that a contractor is installing 50 duplex receptacles, lighting outlets and switches for lighting control on four different jobs, with each having one of the four different installation situations, beginning with the Group 1 working conditions and ending with the situations in Group 4.

Group 1: This situation requires the least labor. All work areas are open and access is unrestricted. An example of this would be a new home with exposed joists and stud walls, allowing easy access for hanging boxes and easy routing of cable. In an existing home, an example might be an unfinished attic with good headroom, an unfinished basement, or a new addition. See Figure 1-1.

Group 2: This situation includes wiring in areas that are partially accessible but that require minor fishing of cable in concealed partitions. For example, installing a switch leg for a lighting outlet in a finished room. The area above the finished ceiling is an accessible attic space. The joists can easily be notched and drilled to accept the new wiring. However, the short distance from the attic space to the switch box in the finished room below is not completely accessible. Consequently, a hole will have to be cut in the wall of the finished room below for the new switch box, and then a cable will have to be fished from the attic space to the wall opening. The same is true for a duplex receptacle installed in the same type of area. Both of these situations are shown in Figure 1-2. If there are fire-stops or diagonal

Figure 1-1: Open wiring in unfinished basements, even if in existing homes, is classified as Group 1 installations.

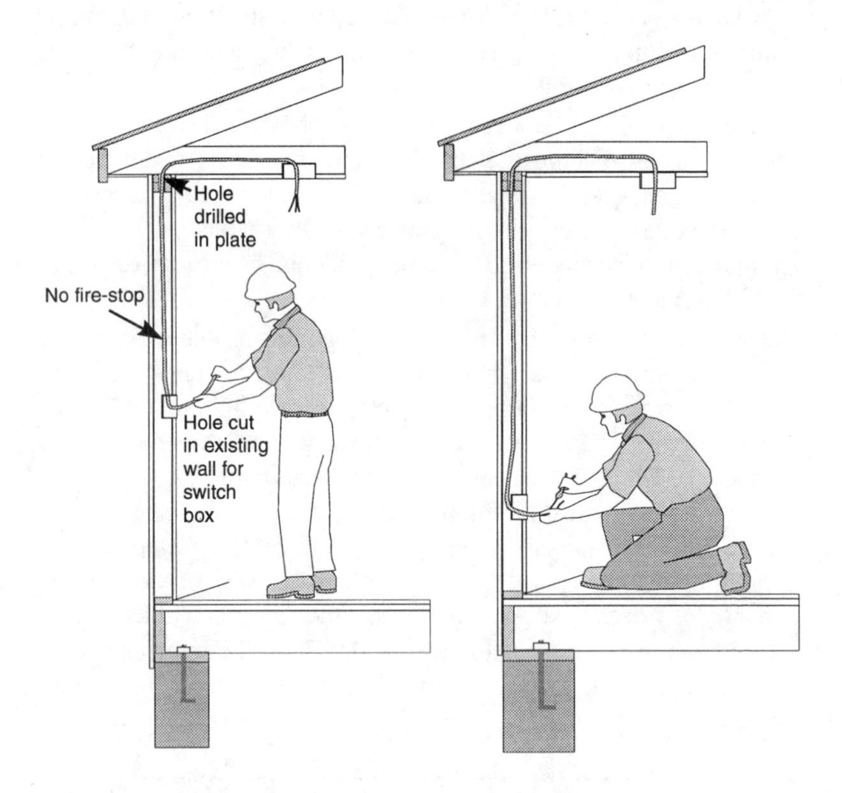

Figure 1-2: Examples of Group 2 installation situations.

bracing in the partition (Figure 1-3), the installation will become more difficult and will then fall into Group 3 installations.

Surface raceways tapped from existing outlets and requiring some changes of direction also fall into the Group 2 category.

Group 3: This installation group covers concealed wiring in partially inaccessible or restricted spaces. For example, working in a restricted crawl space under a house will increase labor time significantly.

Other installation situations that fall into this category include:

- Notching of fire-stops or diagonal bracing to get cables into finished wall spaces.

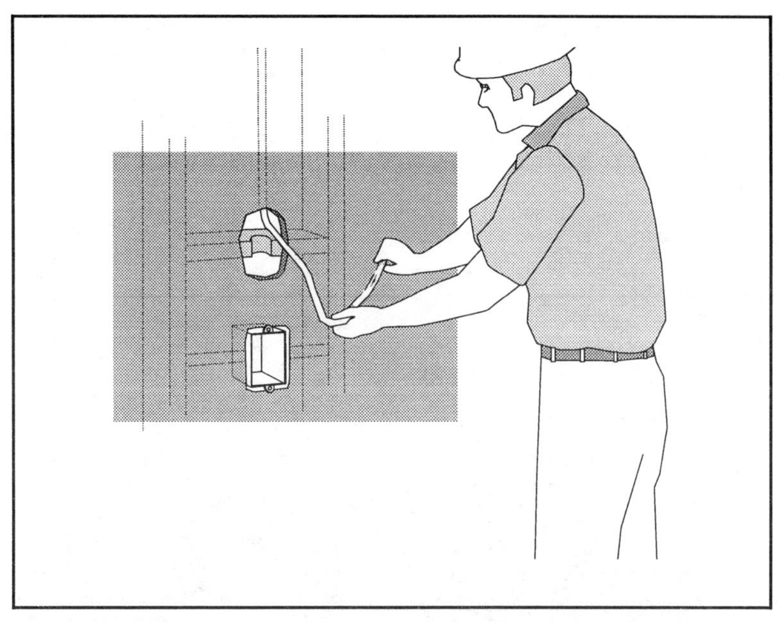

Figure 1-3: If fire-stops or diagonal bracing are encountered, an otherwise Group 2 installation becomes a Group 3 installation.

- Installing wiring on masonry walls where furring strips have been applied.

- Installing wiring in attics or basements where both horizontal and vertical surfaces have been closed in — requiring considerable "fishing" to and from outlets. See Figure 1-4 on the next page.

Group 4: This situation requires the most labor. It includes wiring through masonry walls that have to be cut through using a rotary or star drill and hammer, removal of flooring to route cable, removal of baseboards and door or window trim to route new wiring behind them, and the cutting and patching of finished surfaces to conceal new wiring.

It is likely that more than one installation situation will be found on the same project. For instance, a house may have solid masonry outside walls and frame inside walls or partitions; there may be an unfinished basement under part of the house and only a small crawl

Figure 1-4: Installing wiring in attics or basements where both horizontal and vertical surfaces have been closed in fall into Group 3 installations.

space under the other part. It is the responsibility of the estimator to take this into account when classifying the schedule of outlets or wiring according to groups.

Calculating Prime Cost

Material Costs: Material costs for the same type circuit or outlet for residential work will vary depending upon:

- Whether it is new work or a renovation job.
- The type of construction.

- The quality or type preferences of the contractor and the customer.

When listing material prices on the prime cost determination sheets, the following factors must be considered.

Boxes and Supports: The proper type outlet box and support should be used. When using armored or nonmetallic sheathed cable, boxes with built-in cable clamps should be used. In some instances, especially new work, the box or box hanger may be nailed to the stud. In most modernization work, some other type box support, depending upon the installation situation and the preference of the contractor, must be used. Boxes to which fixtures are to be attached should have fixture studs and require firm support.

Length of Runs: The footage of conduit, wire and cable required will vary for each installation situation group. The shortest run will be in new or open-frame construction, Group 1, and the longest in Group 4. In renovation work, the routing will be that which is easiest to install rather than the most direct.

Conductors per Run: The type and quantities of materials listed in the tables allow two conductors between an outlet and a switch assuming the circuit wires run from outlet to outlet. In some instances, the circuit may go to the switch and then to the outlet, requiring three conductors between the switch and the outlet, in the case of a continuing circuit. This latter method, however, results in an over-all saving of both material and labor and the use of the listed data will not cause the bid to be too low.

A similar instance is a four-way switch installation. The tables show only three conductor cables but in some instances, four conductors may be required.

Quality of Wiring Devices: Switches and receptacles are available in several grades. The price of the grade most commonly used should be inserted in the table to determine the prime cost and the selling price. If the customer desires a higher grade — a mercury switch for instance — the difference in price should be added to the selling price. There will no increase in labor, and if it is believed that additional overhead and profit are justified, the additional allowances can be included as part of the price of device.

Miscellaneous Material: Under the miscellaneous material category, an allowance should be included for all of the small items such as nails, screws, staples, straps, grounding clips, connectors, etc. An allowance for expendable tools, like drill bits, should also be included.

In the case of residential wiring, the amortization of large tools is generally covered in the overhead allowance, but items like hack saw blades, drill bits, etc., should be included as miscellaneous material. These items cost money and if they are not considered as a part of the cost of the job, they must be paid for out of the profits.

Direct Job Expense: The final item of prime cost is direct job expense.

Direct job expense includes items of cost other than material and labor that can be identified and charged directly to the job.

These items include:

- The labor adder (Social Security, unemployment insurance, workmen's compensation insurance, public liability and property damage insurance, association pension and other welfare assessments).
- Sales tax.
- Permit and inspection fees.
- Any other items of cost that can be charged directly to the job.

Do not confuse these items with overhead. Overhead covers only those costs which cannot be identified with any individual job and includes such things as truck expense, general supervision, office expenses, large tool expense, advertising, etc.

Calculating Selling Price

To obtain the selling price, overhead costs and profit are added to prime cost.

Overhead Cost: Surveys indicate that overhead cost for residential work, particularly renovation work, is generally much higher percentage-wise than for other types of work.

To perform the same total volume of business, many more individual jobs must be obtained than in the commercial or industrial field. The amount of paper work per job is not cut down proportionally with the size of the job, and advertising and selling expense are usually much higher.

A thorough analysis of your overhead cost items pertaining to residential work should be made and a percentage of overhead to prime costs determined and applied. A wrong guess usually means less profit, if not a loss.

Profit: An amount of profit that is reasonable and fair to both the customer and the contractor should be added to the total cost to obtain a selling price. Profit is the ultimate end of being in business. Be sure to include it in the selling price.

How To Use The Tables

A sample table for a conventional duplex receptacle is shown in Figure 1-5 on the next page. Another table, complete with calculations, showing how to use the tables in this book is shown in Figure 1-6 on page 23.

- The unit price for material items and the labor rate should be entered in the Unit Price or Rate column.

- Material extensions are then made by multiplying the unit price by the quantity, bearing in mind that a different priced box and support and varying quantities of wiring cable will be required in the different installation situation groups.

- Next, the labor extension is made. The unit is listed for each group. This multiplied by the local labor rate gives the labor cost. The blocks that are not appropriate have been blocked out.

Cost Item	Quantity	Price or Rate	Per	Groups 1	2	3	4
MATERIAL							
Box and support							
Duplex receptacle and plate							
#14-2 w/grd. Type NM Cable							
Miscellaneous							
				$	$	$	$
TOTAL MATERIAL COST							
TOTAL LABOR COST - GROUP 1	0.75 WH		Hr				
GROUP 2	1.10 WH		Hr				
GROUP 3	1.50 WH		Hr				
GROUP 4	1.80 WH		Hr				
DIRECT JOB EXPENSE							
TOTAL PRIME COST				$	$	$	$

Figure 1-5: Blank table without cost items inserted.

Cost Item	Quantity	Price or Rate	Per	Groups			
MATERIAL				1	2	3	4
Box and support	1		E	.90	.90	1.50	1.50
Duplex receptacle and plate	1	1.50		1.50	1.50	1.50	1.50
#14-2 w/grd. Type NM Cable	20-35 ft.		C ft.	3.00	3.75	4.50	5.25
Miscellaneous	Lot			.50	.50	.50	.50
TOTAL MATERIAL COST				$ 5.90	$ 6.65	$ 8.00	$ 8.75
TOTAL LABOR COST - GROUP 1	0.75 WH		Hr	11.25			
GROUP 2	1.10 WH		Hr		16.50		
GROUP 3	1.50 WH		Hr			22.50	
GROUP 4	1.80 WH		Hr				27.00
DIRECT JOB EXPENSE				1.20	1.80	2.40	3.00
TOTAL PRIME COST				$ 18.35	$ 24.95	$ 32.90	$ 38.75

Figure 1-6: Completed table to arrive at prime cost of work.

Prime Cost: The dollar amounts are then added vertically for each group and the totals, which are the prime costs, are entered in the blocks opposite "Total Prime Cost."

Overhead: The prime costs are then entered on the Selling-Price sheets on the line headed "Total Prime Cost" as shown in Figure 1-7. The overhead percentage figure related to prime cost is then entered in the appropriate space on the "Overhead" line and applied to each of the prime cost figures to obtain dollar amount of overhead for each type of outlet or circuit. This dollar amount of overhead is then added to the prime cost to obtain total cost.

Profit: The percentage of profit desired is entered on the "Profit" line and applied to the total cost figures to obtain the dollar amounts of profit. Adding the profit to the total cost will give the selling price. These are figures used to estimate the bid for a given project. A completed table is shown in Figure 1-8.

Reference To The Tables

It is advisable to have a schedule of calculated unit selling prices for quick reference at all times. Much extra work may be obtained on the job site if the contractor or estimator is able to give a quick cost for such work. This may be accomplished in several different ways:

Duplex Receptacle Type NM Cable		**GROUPS**			
		1	**2**	**3**	**4**
Total Prime Cost		$	$	$	$
Overhead	_% of prime cost				
Total Cost		$	$	$	$
Profit	_% of total cost				
Selling Price		$	$	$	$

Figure 1-7: Sample form used for selling price determination.

Duplex Receptacle Type NM Cable		GROUPS			
		1	2	3	4
Total Prime Cost		$18.35	$24.95	$32.90	$38.75
Overhead	30% of prime cost	5.50	7.49	9.87	11.63
Total Cost		$23.85	$32.44	$42.77	$50.38
Profit	25% of total cost	5.96	8.11	10.69	12.60
Selling Price		$29.81	$40.55	$53.46	$62.98

Figure 1-8: Sample of selling price determination.

- Obtain extra copies of this book, one for each estimator and job supervisor. Make sure the information is identical in each. Carry the book with you at all times during work hours.

- Duplicate only the selling prices in this book and keep them in a loose-leaf binder. Again, always have this data available during work hours.

The remaining pages of this book contain worksheets and charts that are designed to enable each individual contractor or estimator to establish their own selling prices for residential electrical work. All popular wiring methods are covered for almost every conceivable application. Forms are also provided for unusual applications.

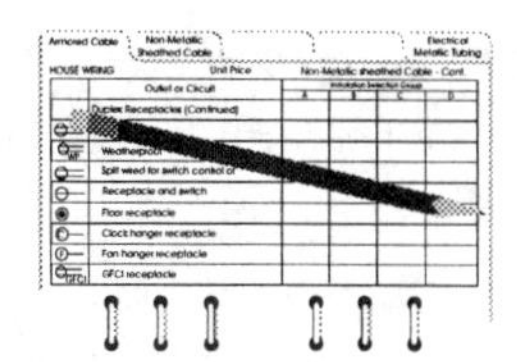

Chapter 2

Sample Estimate

Most new residential construction projects will use a set of working drawings consisting of a site plan, floor plans, elevations, sections, detail drawings, and schedules for the basic framing and construction. Many homes, however, are not large enough to justify the expense of preparing a complete set of detailed electrical drawings and written specifications. Such drawings are usually furnished only on very large homes or tract development projects. Consequently, the electrical contractor will have to perform some design work prior to estimating most projects, and then prepare a complete set of working drawings before the job gets under way.

Sizing The Electric Service

Sometimes it is confusing just which comes first: the layout of the outlets, or the sizing of the electric service. In many cases, the service size (size of main disconnect, panelboard, service conductors, etc.) can be sized using *National Electrical Code® (NEC)* procedures before the outlets are actually located. In other cases, the outlets will have to be laid out first. However, in either case, the service-entrance and panelboard locations will have to be determined before the circuits can be installed — so the contractor or his electricians will know in which direction (and to what points) the circuit homeruns will terminate. Let's take an actual residence and size the electric service according to the latest edition of the *NEC*.

A floor plan of a small residence is shown in Figure 2-1. This building is constructed on a concrete slab with no basement or crawl space. There is an unfinished attic above the living area, and an open carport just outside the kitchen entrance. Appliances include a 12 kVA (12,000 volt-amperes or 12 kilovolt-amperes) electric range and a 4.5 kVA water heater. There is also a washer/dryer (the latter rated at 5.5 kVA) in the utility room. Gas heaters are installed in each room with no electrical requirements. If no other information is furnished, the first step in estimating the cost of the entire project is to calculate the service-entrance size.

General Lighting Loads

General lighting loads are calculated on the basis of *NEC* Table 220-3(b). For residential occupancies, three volt-amperes (watts) per square feet of living space is the figure to use. This includes non-appliance duplex receptacles into which table lights, television, etc. may be connected. Therefore, the area of the building must be calculated first. If the building is under construction, the dimensions can be determined by scaling the working drawings used by the builder. If the residence is an existing building, with no drawings, actual measurements will have to be made on the site.

Using the floor plan of the residence in Figure 2-1 as a guide, an architect's scale is used to measure the longest width of the building (using outside dimensions) which, in this case, is 33 feet. The longest length of the building is 48 feet. These two measurements when multiplied together give (33′ x 48′ =) 1584 square feet of living area. However, there is an open carport on the lower left of the drawing. The carport area will have to be calculated and then deducted from the 1584 square foot figure above to give a true amount of living space. The carport area is 12 feet wide by 19.5 feet long. So, 12 feet by 19.5 feet = 234 square feet. Consequently, the carport area deducted from 1584 square feet leaves (1584 – 234 =) 1350 square feet of living space. See Figure 2-2 on page 32.

When using the square-foot method to determine lighting loads for buildings, *NEC* Section 220-3(b) requires the floor area for each floor

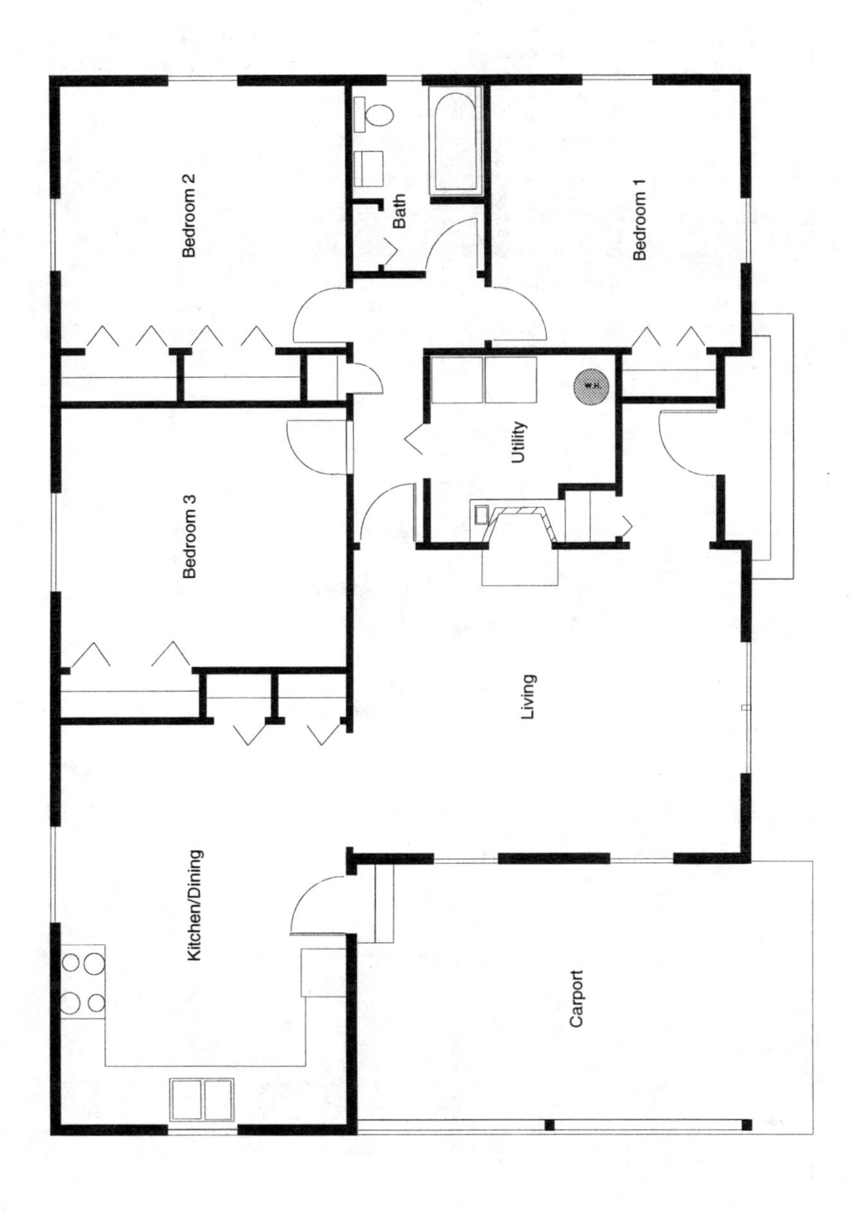

Figure 2-1: Floor plan of a small residence.

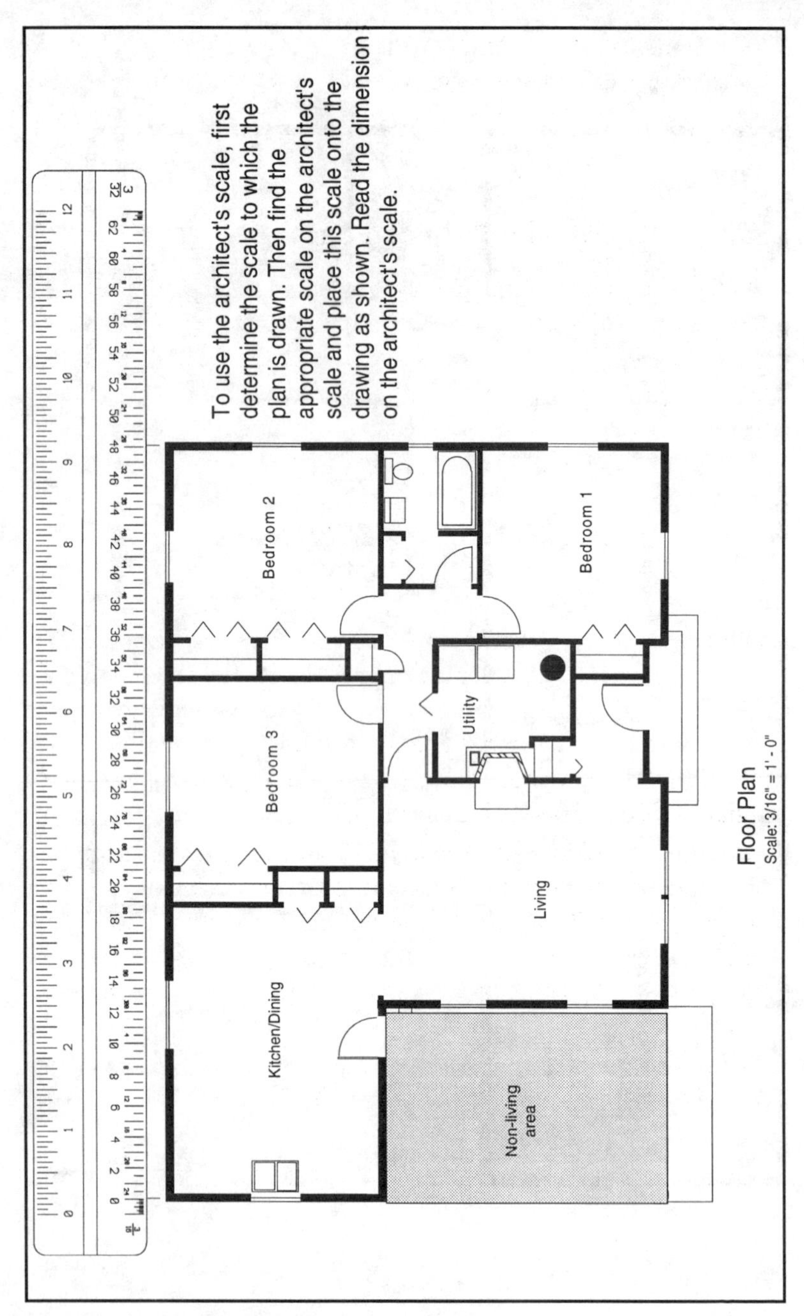

Figure 2-2: Method used to scale a floor plan.

to be computed from the *outside* dimensions. When calculating lighting loads for residential occupancies, the computed floor area must not include open porches, carports, garages, or unused or unfinished spaces not adaptable for future use.

Calculating Electric Service Load

Figure 2-3 on the next page shows a standard calculation worksheet for a single-family dwelling. This form contains numbered blank spaces to be filled in while making the service calculation. Using this worksheet as a guide, we know that the total area of the sample dwelling is 1350 square feet of living space. This figure is entered in the appropriate space (1) and then multiplied by 3 VA for a total general lighting load of 4050 volt-amperes (2).

Small Appliance Loads: *NEC* Section 220-4(b) requires at least two, 120-volt, 20-ampere small appliance circuits to be installed for small appliance loads in the kitchen, dining area, breakfast nook, and similar areas where toasters, coffee makers, etc. will be used. *NEC* Section 220-16 gives further requirements for residential small appliance circuits; that is, each must be rated at 1500 volt-amperes. Since two circuits are used in the sample residence, the number "2" is entered in the appropriate space (3) and then multiplied for a total appliance load of 3000 volt-amperes (4).

Laundry Circuit: *NEC* Section 220-4(c) requires an additional 20-ampere branch circuit to be provided for the exclusive use of the laundry area (5). This circuit must not have any other outlets connected except for the laundry receptacle(s) as required by *NEC* Section 210-52(f). Therefore, enter 1500 (volt-amperes) in space "6" on the form.

Thus far, enough information has been compiled to complete the first portion of the service-calculation form.

General Lighting .... 4050 VA (2)
Small Appliance Load........................ 3000 VA (4)
Laundry Load .... 1500 VA (6)
Total General Lighting, Small
Appliance and Laundry Loads .... 8550 VA (7)

I. General Lighting Load

Type of Load	Calculation	Total VA	*NEC* Reference
Lighting load	(1)______sq. ft. x 3 VA =	(2)______VA	Tbl. 220-3(b)
Small appliance load	(3)____circuits x 1500VA =	(4)______VA	220-16(a)
Laundry load	(5)____circuits x 1500VA =	(6)______VA	220-16(b)
Lighting, small appliance, and laundry	Total VA =	(7)______VA	
Load	**Calculation**	**Demand**	**Total VA**
Lighting, small appliance, and laundry	First 3000VA x	100% =	(8) 3000VA
	(9) Remaining VA x	35% =	(10)_____VA
	Add #8 and #10		(11)_____VA
Total Calculated Load For Lighting, Small Appliances & Laundry			(12)_____VA

II. Large Appliance Loads

Type of Load	Nameplate Rating	Demand	Total VA
Electric range	Not over 12 kVA	Use 8 kVA	(13) 8,000 VA
Clothes dryer	(14)________VA	100%	(15)________VA
Water heater	(16)________VA	100%	(17)________VA
Other appliances	(18)________VA	100%	(19)________VA
Total Large Appliance Load (add items 13, 15, 17 and 19)			(20)________VA
Total Calculated Load (add #12 +#20 =)			(21)________VA

III. Convert VA To Amperes

$$\frac{\textit{Total VA in Box \#21}}{240 \textit{ volts}} = \text{amperes}$$

Figure 2-3: Calculation worksheet for residential service requirements.

Demand Factors

All residential electrical outlets are never used at one time. There may be a rare instance where all lighting fixtures may be on for a short time during the night, but if so, all the small appliances, all burners on the electric range, water heater, furnace, dryer, washer, and the numerous receptacles throughout the house will never be used simultaneously. Knowing this, the *NEC* allows a diversity or demand factor in sizing electric services.

First 3000 VA is rated at 100% .... 3000 VA (8)
The remaining 5550 VA (9) may
be rated at 35% (demand factor).
Therefore, 5550 x .35 = ... 1942.5 VA (10)
Net General Lighting and Small
Appliance load................................ 4942.5 VA (11)

This number (4942.5 VA) is entered in space No. 12 on the form.

The electric range, water heater, and clothes dryer must now be considered in the service calculation. Although it has been determined that the nameplate rating of the electric range is 12 kVA, seldom will every burner be on high at once. Nor will the oven remain on all the time during cooking. When the oven reaches the temperature set on the oven control, the thermostat shuts off the power until the oven cools down. Again, the *NEC* allows a diversity or demand factor.

When one electric range is installed and the nameplate rating is not over 12 kVA, NEC Table 220-19 allows a demand factor resulting in a total rating of 8 kVA. Therefore, 8 kVA may be used in the service calculations instead of the nameplate rating of 12 kVA. The electric clothes dryer and water heater, however, must be calculated at 100% when using this method to calculate residential electric services. The total large appliance load is entered in space No. 20 in the form. Here is a summary of the service calculation thus far:

Net General Lighting and
Small Appliance Load..................... 4942.5 VA (11) & (12)

Electric Range (using demand).............8000 VA
Clothes Dryer ....5500 VA (14) & (15)
Water Heater ....4500 VA (16) & (17)

Total Load ....22942.5 VA (21)

Required Service Size

The conventional electric service for residential use is 120/240-volt, 3-wire, single-phase. Services are sized in amperes and when the volt-amperes are known on single-phase services, amperes may be found by dividing the (highest) voltage into the total volt-amperes. In the case of the sample residence calculations, the computations are as follows:

$$22942.5(\text{VA}) \div 240\ (\text{volts}) = 95.6\ \text{amperes}\ (22)$$

When the net computed load exceeds 10 kVA, or there are six or more two-wire branch circuits, the minimum size service conductors and panelboard must be 100 amperes as required in *NEC* Section 230-42.

The service-entrance conductors have now been calculated and must be rated at 100 amperes. See Figure 2-4 for a completed calculation form for the residence in question.

In *NEC* "Notes to Ampacity Tables (No. 3)" that follow *NEC* Table 310-19, special consideration is given 120/240-volt, 3-wire, single-phase residential services. Conductor sizes are shown in the table that follows these *NEC* notes. Reference to this table shows that the *NEC* allows a No. 4 AWG copper or No. 2 AWG aluminum or copper-clad aluminum for a 100-ampere service. Futhermore, the *NEC* states that the grounded or neutral conductor must never be sized less than two wire sizes smaller than the ungrounded conductors. Therefore, the grounded conductor for a 100-ampere service must not be smaller than No. 8 AWG copper or No. 4 AWG aluminum.

When sizing the grounded conductor for services, provisions stated in *NEC* Section 215-2, 220-22, and 230-42 must be followed, along with other applicable Sections.

I. General Lighting Load

Type of Load	Calculation	Total VA	*NEC* Reference
Lighting load	(1)1350 sq. ft. x 3 VA =	(2)4050 VA	Tbl. 220-3(b)
Small appliance load	(3)2 circuits x 1500VA =	(4)3000 VA	220-16(a)
Laundry load	(5)1 circuits x 1500VA =	(6)1500 VA	220-16(b)
Lighting, small appliance, and laundry	Total VA =	(7)8550 VA	
Load	**Calculation**	**Demand**	**Total VA**
Lighting, small appliance, and laundry	First 3000VA x	100% =	(8) 3000VA
	(9) Remaining VA 5550 x	35% =	(10)1942.5VA
	Add #8 and #10		(11)4942.5VA
Total Calculated Load For Lighting, Small Appliances & Laundry			(12)4942.5 VA

II. Large Appliance Loads

Type of Load	Nameplate Rating	Demand	Total VA
Electric range	Not over 12 kVA	Use 8 kVA	(13) 8,000 VA
Clothes dryer	(14)5500 VA	100%	(15)5500 VA
Water heater	(16)4500 VA	100%	(17)4500 VA
Other appliances	(18)________VA	100%	(19)________VA
Total Large Appliance Load (add items 13, 15, 17 and 19)			(20)18,000 VA
Total Calculated Load (add #12 +#20 =)			(21)22,942.5 VA

III. Convert VA To Amperes

$$\frac{22{,}942.5}{240\ \mathit{volts}} = 95.6 \text{ amperes}$$

Figure 2-4: Completed service-entrance calculation worksheet.

Estimating Cost Of Service-Entrance

The information obtained thus far should be noted and recorded on some type of estimating form or note paper. One type of estimating form that may be used appears in Figure 2-5.

Residential Electrical Estimating Form

ITEM	SIZE OR TYPE	TOTAL QTY.	UNIT PRICE	TOTAL PRICE
Service-Entrance	100 A	1		
Panelboard				
Duplex Receptacles	15A			
Duplex Receptacles	20A			
Duplex Receptacles	W.P.			
Duplex Receptacles	GFCI			
Lighting Outlets, Ceiling				
Lighting Outlets, Wall				
Lighting Outlets, Other				
Single-Pole Wall Switch				
Three-Way Switches				
Four-Way Switches				
Electric Range				
Clothes Dryer				
Water Heater				
Lighting Fixtures				
Miscellaneous				

FOR:

Name____________________

Address____________________

City____________________

State__________ Zip_______

Phone____________________

Total Net Price	
Sales Tax	
Total Price	
Deposit	
Balance	

Figure 2-5: One type of estimating form that may be used for unit pricing.

We could price the service-entrance at this time, using the selling-price tables in this book. However, it is usually best to make a complete take-off of all items first; then, the pricing may be done for all items at one time. Consequently, let's continue with the estimate for the residence in question.

Sizing The Panelboard

Notice that the second item in the estimating form in Figure 2-5 is the panelboard, sometimes referred to as a load center in residential wiring.

The *NEC* requires that each ungrounded conductor in all electrical circuits must be provided with overcurrent protection — either in the form of fuses or circuit breakers. If more than six such devices are used, a means of disconnecting the entire service must be provided, using either a main disconnect switch or a main circuit breaker.

To calculate the number of overcurrent protective devices required for the sample residence, let's look at the general lighting load first. Since there is a total general lighting load of 4050 volt-amperes, this figure can be divided by 120 volts (Ohm's Law states amperes = VA/V) which equals:

$$4050 \div 120 = 33.75$$

Either 15- or 20-ampere circuits may be used for the lighting load. However, let's use 15-ampere circuits in this case. Consequently, three 15-ampere, 1-pole circuit breakers will be required for the total lighting load.

In addition to the lighting circuits, the sample residence will require a minimum of two 20-ampere circuits for the small-appliance load and one 20-ampere circuit for the laundry area. Thus far, we can count the following branch circuits:

General Lighting Load .... Three 15-amp circuits
Small-Appliance Load .... Two 20-ampere circuits
Laundry Load .... One 20-ampere circuit

Total............................. Six 1-Pole Breakers

Most load centers and panelboards are provided with an even number of circuit breaker (cb) spaces; that is, 4, 6, 8, 10, etc. However, before the panelboard can be selected for the sample residence, space must be provided for the 240-volt loads — each requiring a 2-pole circuit breaker or fuse holder.

The rating of overcurrent devices for the 240-volt loads are sized by dividing the voltage into the demand volt-amperes. For example, since the demand load of the electric range is 8 kVA, 8000 volt-amperes divided by 240 volts equals 33.3 amperes. The closest standard overcurrent device is 40 amperes; this will be the size used — a 40-ampere, 2-pole circuit breaker. The remaining 240-volt circuits are calculated in a similar fashion, resulting in the following:

- Clothes dryer — one 30-ampere, 2-pole circuit breaker
- Water heater — one 30-ampere, 2-pole circuit breaker

These three appliances will therefore require an additional (2 poles x three appliances =) six spaces in the panelboard. Adding these six spaces to the six required for the general lighting and small-appliance loads requires at least a 12-space panelboard (load center) to handle the circuits in our sample residence that have been addressed thus far.

Ground-Fault Circuit-Interrupters

Although the layout of the various outlets in the sample residence has not yet been discussed, the experienced electrician knows that circuits providing power to certain areas of the home require ground-fault circuit-interrupters (GFCIs) to be installed for additional protection of people using these circuits. Such areas include:

- All outside receptacles
- Receptacles used in bathrooms
- Receptacles located in residential garages
- Receptacles located in unfinished basements

- Receptacles located in crawl spaces

- Receptacles located within six feet of a kitchen or bar sink.

Since there is no basement or crawl space in the sample residence, these two areas do not apply to this project. However, there is a bathroom and kitchen. Furthermore, outside receptacles will be provided. Therefore, at least one small-appliance (20-ampere) circuit will be provided with GFCI protection along with one circuit supplying outdoor receptacles. The bathroom receptacle can be connected to the GFCI circuit supplying the outdoor receptacles, or a GFCI receptacle connected to a conventional circuit can be used.

GFCI circuit breakers require one space in the load center or panelboard — the same as a 1-pole circuit breaker. Consequently, at least two more spaces must be added to the 12 spaces obtained in previous calculations; the panelboard spaces now total 14.

We could get by with a 14-space circuit-breaker load center with a 100-ampere main breaker. However, it is always better to provide a few extra spaces for future use. Some local ordinances require 20% extra space in any residential load center for additional circuits that may be added later. The spare circuit breakers do not necessarily have to be installed, but space should be provided for them. This puts the size of the load center for the house in question up to 18 spaces, providing four extra spaces for future use. This information should be entered on the estimating form or note paper.

Estimating Cost Of Receptacles

The electrical contractor should make a sketch of the floor plan and place receptacles at locations to comply with the *NEC* or the requirements of the owner/architect. A floor plan showing one possible receptacle layout for the sample residence is shown in Figure 2-6 on the next page. All receptacle outlets should be counted and segregated into the appropriate category; that is, similar to the following listing:

- Conventional 15-ampere wall receptacles

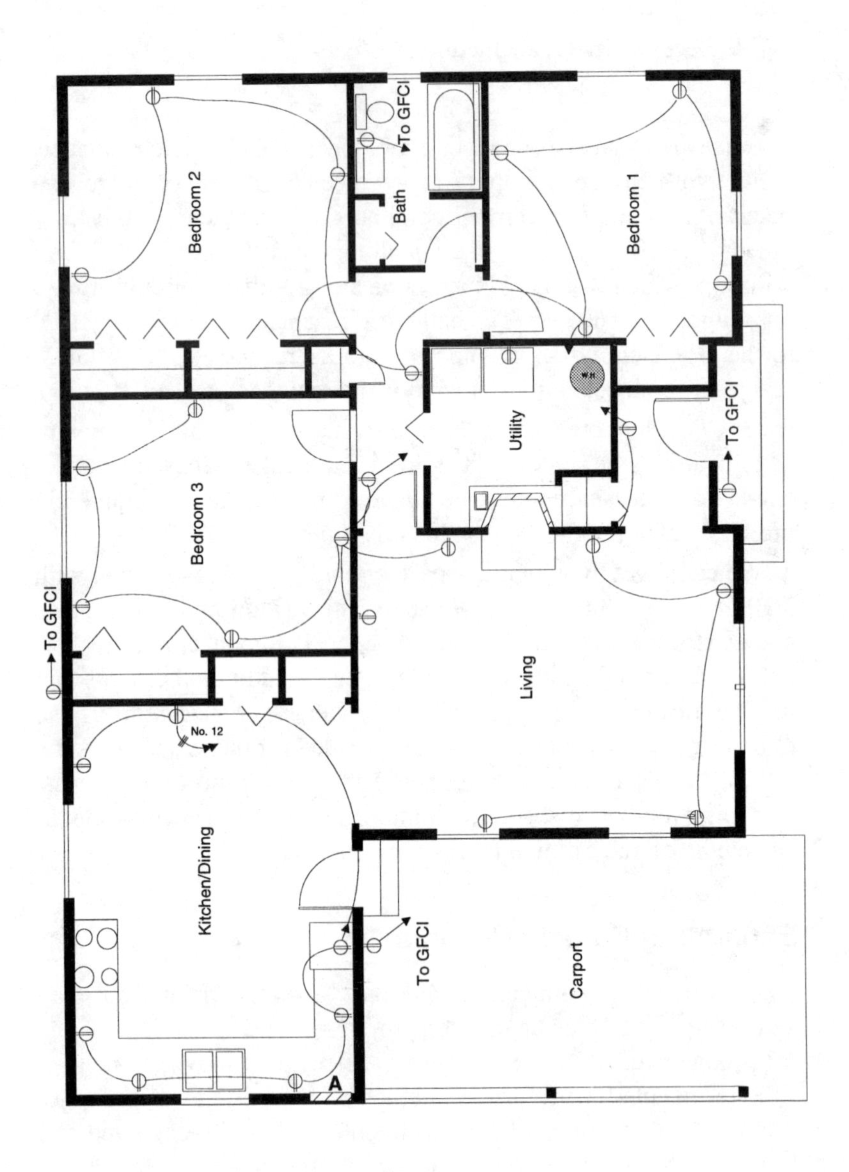

Figure 2-6: One possible receptacle layout for the sample residence.

- Small-appliance, 20-ampere receptacles
- Weatherproof receptacles
- GFCI receptacles

Referring again to Figure 2-6, note that the above types of receptacles number as follows:

- 21 15-ampere receptacles
- 2 20-ampere receptacles
- 5 20-ampere receptacles on GFCI circuit
- 4 15-ampere receptacles on GFCI circuit

The quantity and types of these receptacles should be recorded on the estimate form or else on a note pad before continuing with the estimate. Note also that three of the 15-ampere GFCI receptacles will require weatherproof covers since they are installed outdoors.

Lighting Outlets

When using the unit-pricing method for estimating electrical work, the lighting outlets are divided into two separate categories; that is, one phase of the installation consists of installing only the outlet boxes and related wiring, while the second phase covers the cost of the lighting fixtures. Since the cost of lighting fixtures can vary greatly, usually a "fixture allowance" is stipulated in the contract for residential projects. In the sample residence, $500 will be allocated for the cost of lighting fixtures. The actual installation of the fixtures, and related wiring, however, will have to appear on the estimate form.

A lighting floor plan of the sample residence is shown in Figure 2-7 on the next page. Note that each lighting outlet and its control are indicated by symbol, and the branch circuits and switch legs are shown via lines — indicating exactly how each lighting fixture is to be controlled.

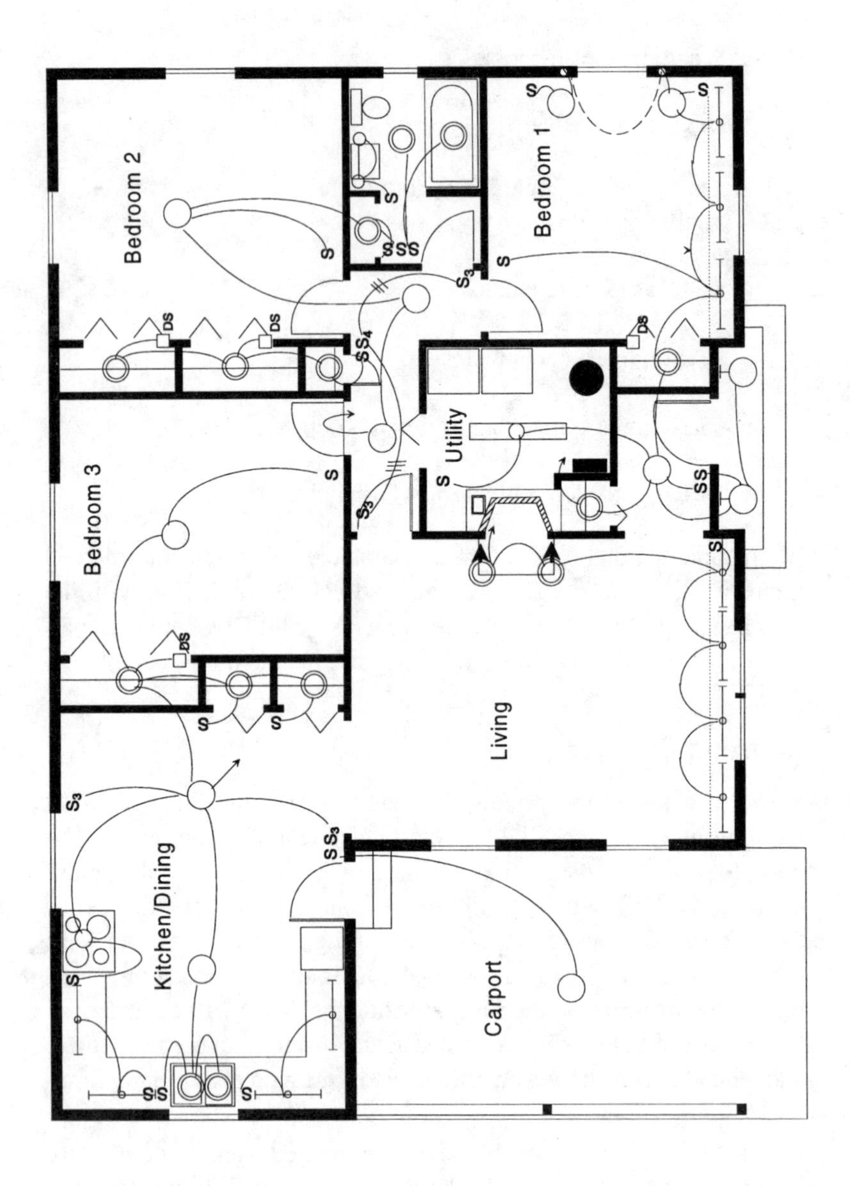

Figure 2-7: Floor plan of lighting layout for sample residence.

Lighting-Fixture Schedule		
Symbol	**Description**	**Number**
	Surface-mounted incandescent fixture	10
	Recessed incandescent fixture	13
	Surface-mounted incandescent wall fixture	2
	Recessed accent lights	2
	Surface-mounted fluorescent strip lighting	11
	Surface-mounted fluorescent fixture	1

Figure 2-8: Lighting-fixture schedule for the sample residence.

The types and number of lighting fixtures shown on the floor plan are shown in the lighting-fixture schedule in Figure 2-8. In general, there are five different types:

- Surface-mounted incandescent
- Wall-mounted incandescent
- Recessed incandescent
- Surface-mounted fluorescent strip lights
- Surface-mounted fluorescent with diffuser

Although five different types of lighting fixtures are involved, only three different installation situations will be used:

- Surface-mounted incandescent and fluorescent fixtures
- Wall-mounted incandescent fixtures
- Recessed incandescent fixtures

All lighting fixtures are listed on the estimate form by category and then a count of the wall switches is in order. The type and number of switches are also added to the estimate sheet.

- 18 Single-pole switches
- 4 Three-way switches
- 1 Four-way switch
- 4 Door switches

At this point, all major items should be accounted for and recorded on the estimating form as shown in Figure 2-9. The project is now ready for pricing and finalizing, using the unit-price tables in this book.

Pricing The Items

When reviewing the items listed on the estimating form in Figure 2-9, note that only a brief description of the items is needed. Material and labor quantities have already been addressed in the unit-price tables found throughout this book. Consequently, all that is necessary is to find the appropriate table and enter the unit price in the designated column on the estimating form in Figure 2-9. Extending the unit price to correspond with the number of each item gives a total price for that particular category or item. Once all items have been addressed, the total-price column is totaled to obtain a total net price, or the actual net selling price that is presented to the customer.

For example, let's take the first item listed in the Residential Electrical Estimating Form; that is, *service-entrance.* Since it has already been determined that this project will require a 100-ampere service-entrance, turn to Chapter 5 in this book and look under the selling-price charts for 100-Ampere Service-Entrance, using aluminum Type S.E. cable. Since this project is new construction and no unusual labor situations are involved, this service-entrance will fall under the Group 1 column, which requires the least amount of labor of the four groups involved. Scan across the column until the selling price is found; enter this price in the unit price column on the form.

Residential Electrical Estimating Form

ITEM	SIZE OR TYPE	TOTAL QTY.	UNIT PRICE	TOTAL PRICE
Service-Entrance	100 A	1		
Panelboard, 100A Main	18 spaces	1		
Duplex Receptacles	15A	21		
Duplex Receptacles	20A	2		
Duplex Receptacles, GFCI	20A	5		
Duplex Receptacles, GFCI	15A	4		
Lighting Outlets, Ceiling	Surface	22		
Lighting Outlets, Wall	Surface	2		
Lighting Outlets, Ceiling	Recessed	15		
Single-Pole Wall Switch		18		
Three-Way Switches		4		
Four-Way Switches		1		
Door Switches		4		
Electric Range	40A			
Clothes Dryer	30A			
Water Heater	30A			
Lighting Fixtures	Allowance			$500.00
Miscellaneous				

FOR:

Name____________________

Address__________________

City_____________________

State___________Zip_______

Phone___________________

Total Net Price	
Sales Tax	
Total Price	
Deposit	
Balance	

Figure 2-9: Estimate form with all major items listed.

100-Ampere Service Type SE-AL Cable	120/240 Volts	GROUPS			
		1	2	3	4
Total Prime Cost		$130.00	$170.00	$210.00	$250.00
Overhead	30 % of prime cost	39.00	51.00	63.00	75.00
Total Cost		$169.00	$221.00	$273.00	$325.00
Profit	25 % of total cost	42.25	55.25	68.25	81.25
Selling Price		$211.25	$276.25	$341.25	$406.25

Figure 2-10: Typical selling-price chart for a 100-ampere service.

A typical completed selling-price form may appear as shown in Figure 2-10. Note that this contractor's price for a 100-ampere service, using aluminum Type S.E. cable, under Group 1 installation situation, is $211.25. This amount is entered in the Unit Price column on the form and then extended to obtain a total price. Since only one 100-ampere service-entrance is involved, the total price is the same; that is, $211.25 x 1 = $211.25.

After contractors and estimators gain some experience with this unit-pricing method, many prefer to use a simplified selling-price chart for finalizing a bid. A simplified chart appears in Figure 2-11. Note that the final selling prices for the various installation groups are exactly the same as those in Figure 2-10. However, by omitting the *Total Prime Cost*, *Overhead*, and *Profit* rows, the charts are easier to read during the finalizing process.

100-Ampere Service Type SE-AL Cable	120/240 Volts	GROUPS			
		1	2	3	4
Selling Price		$211.25	$276.25	$341.25	$406.25

Figure 2-11: Simplified selling-price chart.

Residential Electrical Estimating Form

ITEM	SIZE OR TYPE	TOTAL QTY.	UNIT PRICE	TOTAL PRICE
Service-Entrance	100 A	1	$211.25	$211.25
Panelboard, 100A Main	18 spaces	1	250.00	250.00
Duplex Receptacles	15A	21	25.00	525.00
Duplex Receptacles	20A	2	30.00	60.00
Duplex Receptacles, GFCI	20A	5	40.00	200.00
Duplex Receptacles, GFCI	15A	4	35.00	140.00
Lighting Outlets, Ceiling	Surface	22	20.00	440.00
Lighting Outlets, Wall	Surface	2	20.00	40.00
Lighting Outlets, Ceiling	Recessed	15	25.00	375.00
Single-Pole Wall Switch		18	22.00	396.00
Three-Way Switches		4	37.00	148.00
Four-Way Switches		1	45.00	45.00
Door Switches		4	40.00	160.00
Electric Range	40A	1	80.00	80.00
Clothes Dryer	30A	1	70.00	70.00
Water Heater	30A	1	45.00	45.00
Lighting Fixtures	Allowance			$500.00
Miscellaneous	Inspection		75.00	75.00

FOR:

Name____________________

Address____________________

City____________________

State__________Zip_______

Phone____________________

Total Net Price	$3760.25
Sales Tax	112.80
Total Price	$3873.05
Deposit	
Balance	

Figure 2-12: Completed residential estimating form.

The remaining items listed in the form (Figure 2-12) are priced in a similar way. Take the 100-ampere, 18-space panelboard with a 100-ampere main circuit breaker next. Find the selling price in the selling-price charts at the end of Chapter 5.

Continue with the various types of receptacles, lighting outlets, switches, and finally the appliance circuits. Note that an amount of $75.00 has been entered in the Miscellaneous row for an electrical inspection fee. Read the Total Net Price in the appropriate column. After adding any sales tax, the bid for the sample residence is complete.

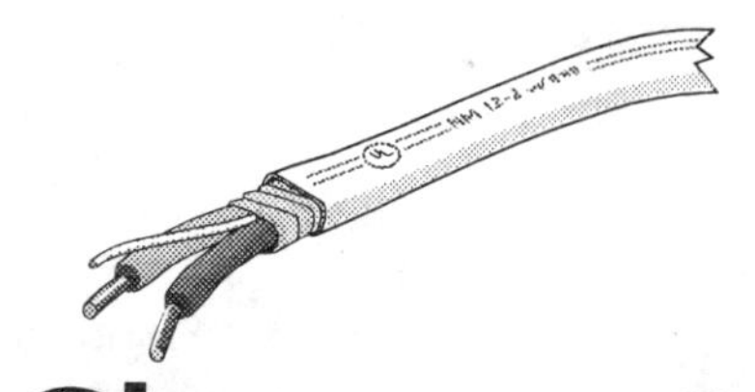

Chapter 3

Nonmetallic-Sheathed Cable

Nonmetallic-sheathed cable (Type NM) is manufactured in two or three-wire assemblies, and with varying sizes of conductors. In both two- and three-wire cables, conductors are color-coded: one conductor is black while the other is white in two-wire cable; in three-wire cable, the additional conductor is red. Both types will also have a grounding conductor which is usually bare, but is sometimes covered with a green plastic insulation — depending upon the manufacturer. The jacket or covering consists of rubber, plastic, or fiber. Most brands also have markings on this jacket giving the manufacturer's name or trademark, the wire size, and the number of conductors. For example, "NM 12-2 W/GRD" indicates that the jacket contains two No. 12 AWG conductors along with a grounding wire; "NM 12-3 W/GRD" indicates three conductors plus a grounding wire. This type of cable may be concealed in the framework of buildings, or in some instances, may be run exposed on the building surfaces. It may not be used in any building exceeding three floors above grade; as a service-entrance cable; in commercial garages having hazardous locations; in theaters and similar locations; places of assembly; in motion picture studios; in storage battery rooms; in hoistways; embedded in poured concrete, or aggregate; or in any hazardous location except as otherwise permitted by the *NEC*. Nonmetallic-sheathed cable is frequently referred to as "Romex" on the job. See Figure 3-1.

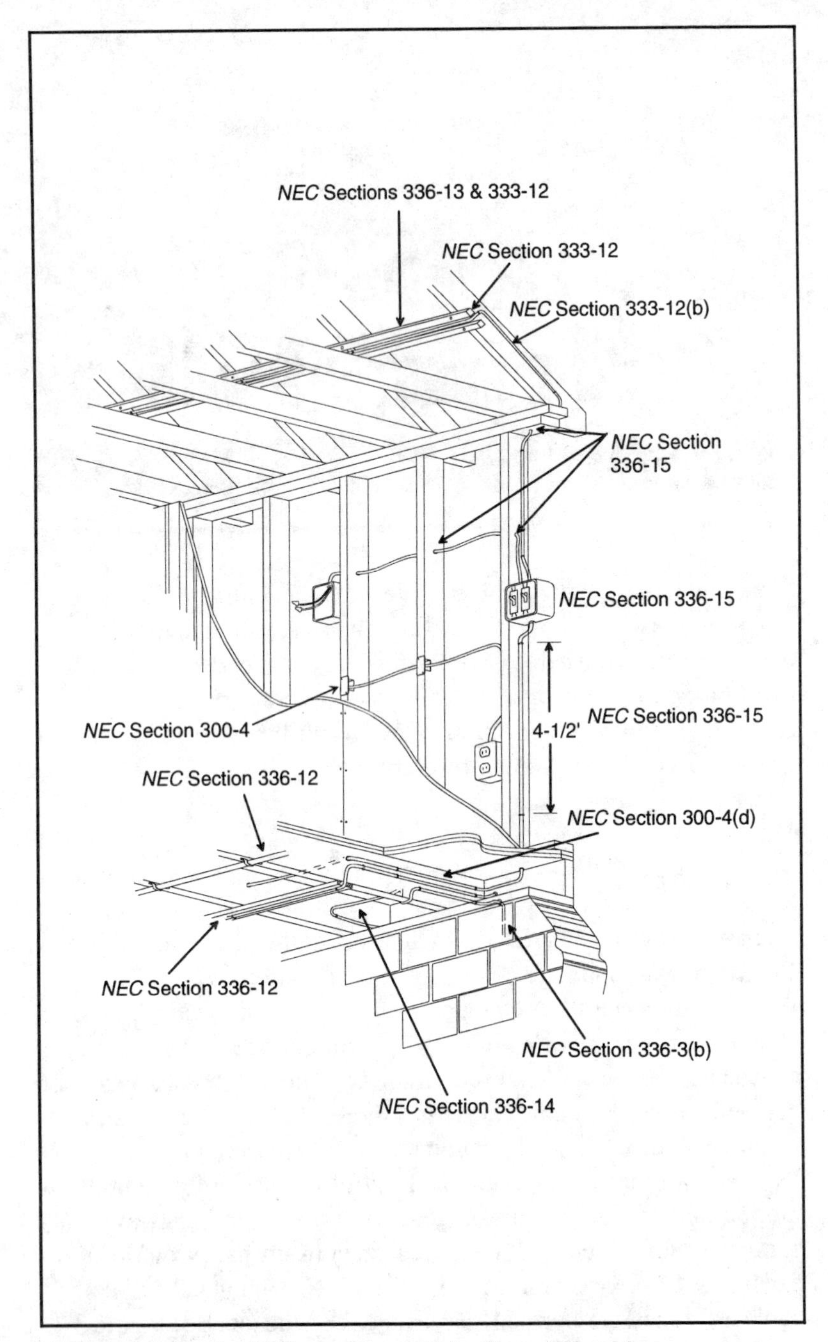

Figure 3-1: Basic *NEC* requirements for Type NM cable.

Surface-Mounted Lighting Outlet

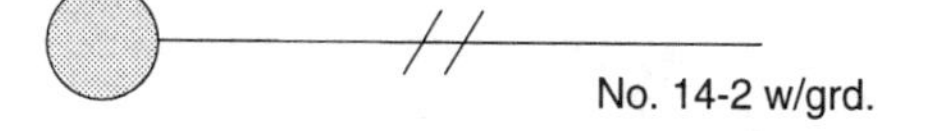

Cost Item	Quantity	Price or Rate	Per	Installation Groups			
MATERIAL				1	2	3	4
Box and support	1		E				
#14-2 w/grd. Type NM cable	20 - 30 ft.		C Ft.				
Miscellaneous	Lot						
TOTAL MATERIAL COST							
TOTAL LABOR COST - GROUP 1	0.35 WH		Hr				
GROUP 2	0.50 WH		Hr				
GROUP 3	0.65 WH		Hr				
GROUP 4	0.80 WH		Hr				
DIRECT JOB EXPENSE							
TOTAL PRIME COST				$	$	$	$

Surface-Mounted Lighting Outlet

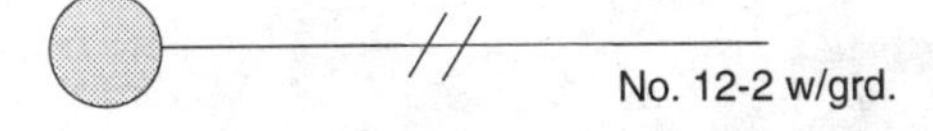

Cost Item	Quantity	Price or Rate	Per	Installation Groups 1	2	3	4
MATERIAL							
Box and support	1		E				
#12-2 w/grd. Type NM cable	20 - 30 ft.		C Ft.				
Miscellaneous	Lot						
TOTAL MATERIAL COST							
TOTAL LABOR COST - GROUP 1	0.40 WH		Hr				
GROUP 2	0.55 WH		Hr				
GROUP 3	0.70 WH		Hr				
GROUP 4	0.85 WH		Hr				
DIRECT JOB EXPENSE							
TOTAL PRIME COST				$	$	$	$

Recessed Lighting Outlet

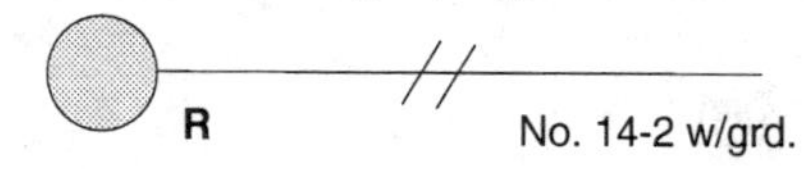

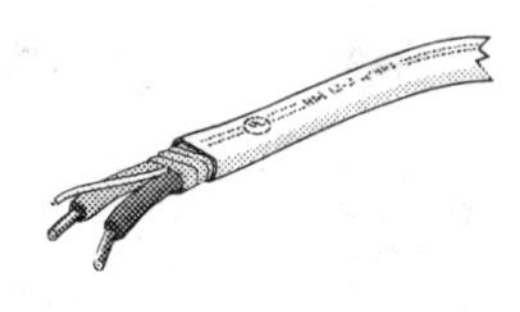

Cost Item	Quantity	Price or Rate	Per	Installation Groups 1	2	3	4
MATERIAL							
Cable connector	1		E				
#14-2 w/grd. Type NM cable	20 - 30 ft.		C Ft.				
Miscellaneous	Lot						
TOTAL MATERIAL COST							
TOTAL LABOR COST - GROUP 1	0.45 WH		Hr				
GROUP 2	0.70 WH		Hr				
GROUP 3	0.90 WH		Hr				
GROUP 4	1.20 WH		Hr				
DIRECT JOB EXPENSE							
TOTAL PRIME COST				$	$	$	$

Recessed Lighting Outlet

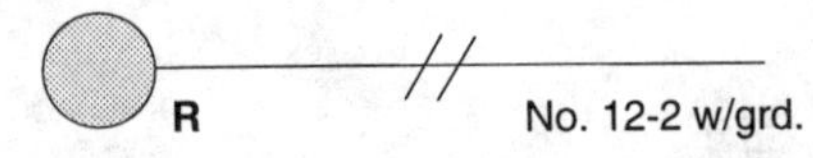

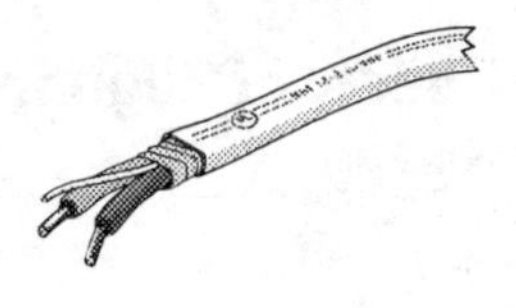

Cost Item	Quantity	Price or Rate	Per	Installation Groups			
				1	2	3	4
MATERIAL							
Cable connector	1		E				
#12-2 w/grd. Type NM cable	20 - 30 ft.		C Ft.				
Miscellaneous	Lot						
TOTAL MATERIAL COST							
TOTAL LABOR COST - GROUP 1	0.55 WH		Hr				
GROUP 2	0.80 WH		Hr				
GROUP 3	1.00 WH		Hr				
GROUP 4	1.30 WH		Hr				
DIRECT JOB EXPENSE							
TOTAL PRIME COST				$	$	$	$

Wall-Mounted Lighting Outlet

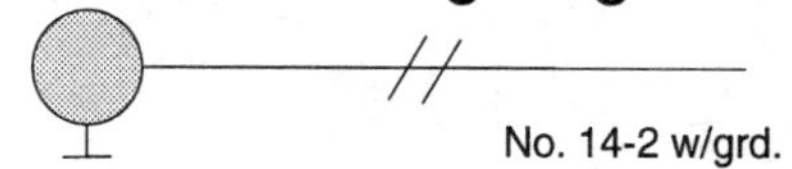

Cost Item	Quantity	Price or Rate	Per	Installation Groups 1	2	3	4
MATERIAL							
Box and support	1		E				
#14-2 w/grd. Type NM cable	20 - 30 ft.		C Ft.				
Miscellaneous	Lot						
TOTAL MATERIAL COST							
TOTAL LABOR COST - GROUP 1	0.30 WH		Hr				
GROUP 2	0.40 WH		Hr				
GROUP 3	0.55 WH		Hr				
GROUP 4	0.70 WH		Hr				
DIRECT JOB EXPENSE							
TOTAL PRIME COST				$	$	$	$

Wall-Mounted Lighting Outlet

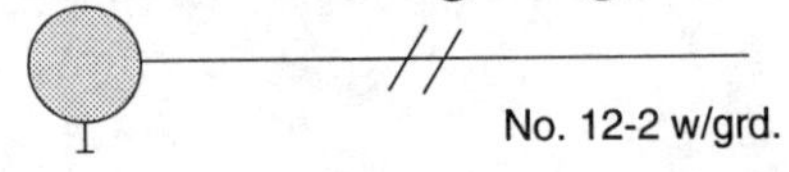

Cost Item	Quantity	Price or Rate	Per	Installation Groups			
				1	2	3	4
MATERIAL							
Box and support	1		E				
#12-2 w/grd. Type NM cable	20 - 30 ft.		C Ft.				
Miscellaneous	Lot						
TOTAL MATERIAL COST							
TOTAL LABOR COST - GROUP 1	0.35 WH		Hr				
GROUP 2	0.50 WH		Hr				
GROUP 3	0.65 WH		Hr				
GROUP 4	0.80 WH		Hr				
DIRECT JOB EXPENSE							
TOTAL PRIME COST				$	$	$	$

Duplex Receptacle – 2-Wire Circuit

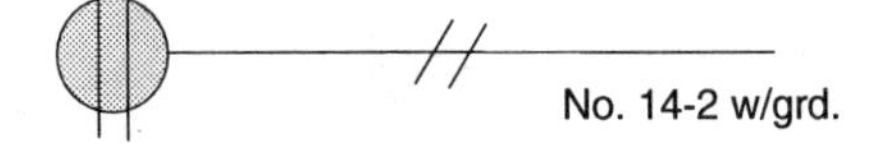

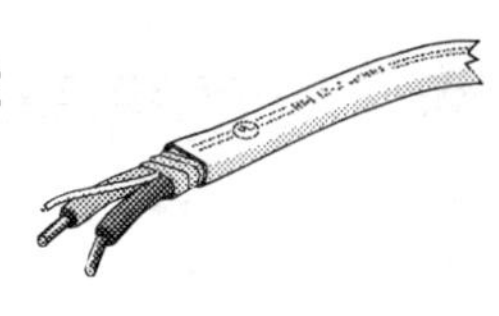

Cost Item	Quantity	Price or Rate	Per	Installation Groups			
				1	2	3	4
MATERIAL							
Box and support	1		E				
#14-2 w/grd. Type NM cable	20 - 30 ft.		C Ft.				
Duplex receptacle and plate	1		E				
Miscellaneous	Lot						
TOTAL MATERIAL COST							
TOTAL LABOR COST - GROUP 1	0.65 WH		Hr				
GROUP 2	1.00 WH		Hr				
GROUP 3	1.35 WH		Hr				
GROUP 4	1.70 WH		Hr				
DIRECT JOB EXPENSE							
TOTAL PRIME COST				$	$	$	$

Duplex Receptacle – 2-Wire Circuit

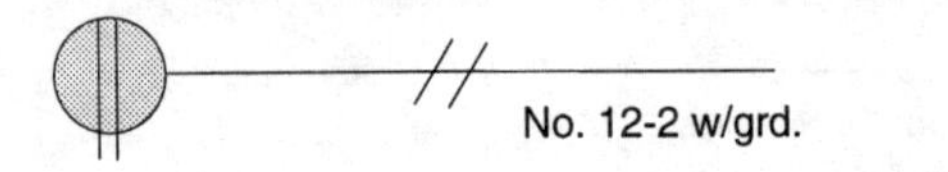

Cost Item	Quantity	Price or Rate	Per	Installation Groups 1	2	3	4
MATERIAL							
Box and support	1		E				
#12-2 w/grd. Type NM cable	20 - 30 ft.		C Ft.				
Duplex receptacle and plate	1		E				
Miscellaneous	Lot						
TOTAL MATERIAL COST							
TOTAL LABOR COST - GROUP 1	0.75 WH		Hr				
GROUP 2	1.10 WH		Hr				
GROUP 3	1.45 WH		Hr				
GROUP 4	1.80 WH		Hr				
DIRECT JOB EXPENSE							
TOTAL PRIME COST				$	$	$	$

Receptacle – 3-Wire Circuit

Cost Item	Quantity	Price or Rate	Per	Installation Groups 1	2	3	4
MATERIAL							
Box and support	1		E				
#14-3 w/grd. Type NM cable	20 - 30 ft.		C Ft.				
Receptacle and plate	1		E				
Miscellaneous	Lot						
TOTAL MATERIAL COST							
TOTAL LABOR COST - GROUP 1	0.55 WH		Hr				
GROUP 2	0.85 WH		Hr				
GROUP 3	1.15 WH		Hr				
GROUP 4	1.45 WH		Hr				
DIRECT JOB EXPENSE							
TOTAL PRIME COST				$	$	$	$

Receptacle – 3-Wire Circuit

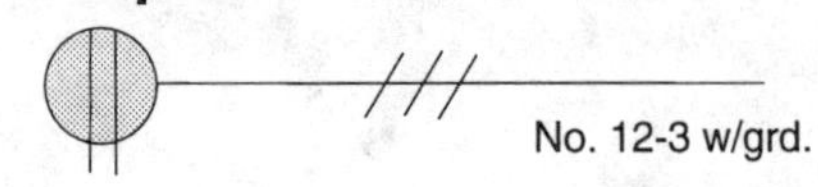

Cost Item	Quantity	Price or Rate	Per	Installation Groups 1	2	3	4
MATERIAL							
Box and support	1		E				
#12-3 w/grd. Type NM cable	20 - 30 ft.		C Ft.				
Receptacle and Plate	1		E				
Miscellaneous	Lot						
TOTAL MATERIAL COST							
TOTAL LABOR COST - GROUP 1	0.65 WH		Hr				
GROUP 2	0.95 WH		Hr				
GROUP 3	1.25 WH		Hr				
GROUP 4	1.50 WH		Hr				
DIRECT JOB EXPENSE							
TOTAL PRIME COST				$	$	$	$

Duplex Receptacle – Split-Wired

Cost Item	Quantity	Price or Rate	Per	Installation Groups			
MATERIAL				1	2	3	4
Box and support	1		E				
#14-3 w/grd. Type NM cable	20 - 30 ft.		C Ft.				
Two-circuit receptacle and plate	1		E				
Miscellaneous	Lot						
TOTAL MATERIAL COST							
TOTAL LABOR COST - GROUP 1	0.60 WH		Hr				
GROUP 2	0.85 WH		Hr				
GROUP 3	1.20 WH		Hr				
GROUP 4	1.55 WH		Hr				
DIRECT JOB EXPENSE							
TOTAL PRIME COST				$	$	$	$

Duplex Receptacle – Split-Wired

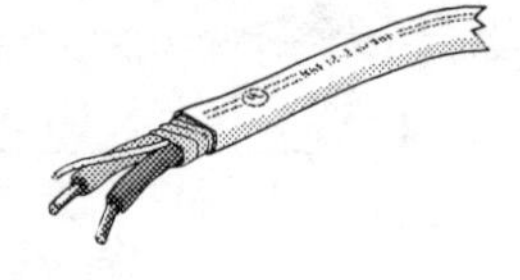

Cost Item	Quantity	Price or Rate	Per	Installation Groups 1	2	3	4
MATERIAL							
Box and support	1		E				
#12-3 w/grd. Type NM cable	20 - 30 ft.		C Ft.				
Two-circuit receptacle and plate	1		E				
Miscellaneous	Lot						
TOTAL MATERIAL COST							
TOTAL LABOR COST - GROUP 1	0.65 WH		Hr				
GROUP 2	0.95 WH		Hr				
GROUP 3	1.25 WH		Hr				
GROUP 4	1.55 WH		Hr				
DIRECT JOB EXPENSE							
TOTAL PRIME COST				$	$	$	$

Duplex Receptacle – Weatherproof

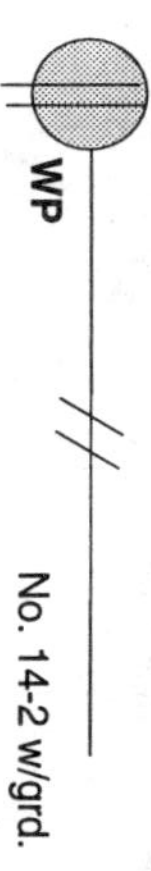

Cost Item	Quantity	Price or Rate	Per	Installation Groups			
MATERIAL				1	2	3	4
Box and support	1		E				
#14-2 w/grd. Type NM cable	20 - 30 ft.		C Ft.				
Receptacle and weatherproof plate	1		E				
Miscellaneous	Lot						
TOTAL MATERIAL COST							
TOTAL LABOR COST - GROUP 1	0.65 WH		Hr				
GROUP 2	1.00 WH		Hr				
GROUP 3	1.35 WH		Hr				
GROUP 4	1.70 WH		Hr				
DIRECT JOB EXPENSE							
TOTAL PRIME COST				$	$	$	$

Receptacle – Weatherproof

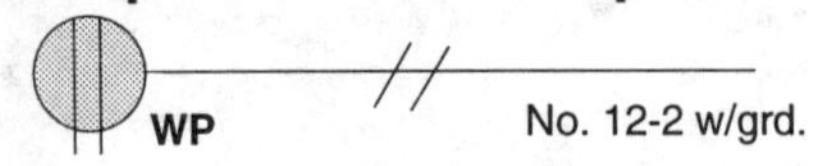

Cost Item	Quantity	Price or Rate	Per	Installation Groups 1	2	3	4
MATERIAL							
Box and support	1		E				
#12-2 w/grd. Type NM cable	20 - 30 ft.		C Ft.				
Receptacle and weatherproof plate	1		E				
Miscellaneous	Lot						
TOTAL MATERIAL COST							
TOTAL LABOR COST - GROUP 1	0.75 WH		Hr				
GROUP 2	1.10 WH		Hr				
GROUP 3	1.45 WH		Hr				
GROUP 4	1.75 WH		Hr				
DIRECT JOB EXPENSE							
TOTAL PRIME COST				$	$	$	$

Duplex Receptacle – GFCI

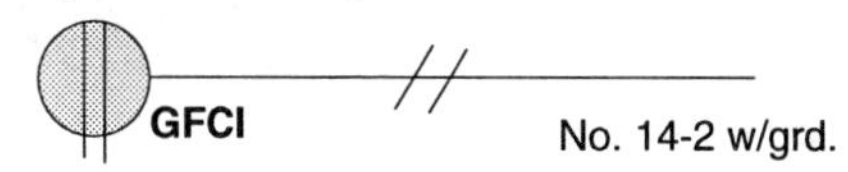

Cost Item	Quantity	Price or Rate	Per	Installation Groups			
				1	2	3	4
MATERIAL							
Box and support	1		E				
#14-2 w/grd. Type NM cable	20 - 30 ft.		C Ft.				
GFCI receptacle and plate or	1		E				
GFCI circuit breaker and regular recept.	1		E				
Miscellaneous	Lot						
TOTAL MATERIAL COST							
TOTAL LABOR COST - GROUP 1	0.65 WH		Hr				
GROUP 2	0.95 WH		Hr				
GROUP 3	1.25 WH		Hr				
GROUP 4	1.55 WH		Hr				
DIRECT JOB EXPENSE							
TOTAL PRIME COST				$	$	$	$

Duplex Receptacle – GFCI

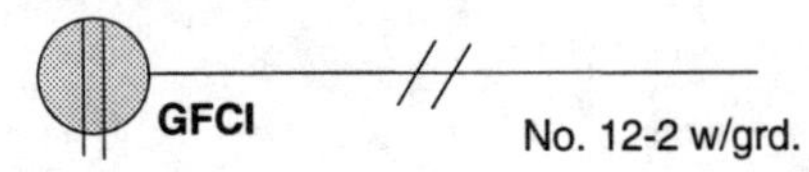

Cost Item	Quantity	Price or Rate	Per	Installation Groups 1	2	3	4
MATERIAL							
Box and support	1		E				
#12-2 w/grd. Type NM cable	20 - 30 ft.		C Ft.				
GFCI receptacle and plate or	1		E				
GFCI circuit breaker and regular recept.	1		E				
Miscellaneous	Lot						
TOTAL MATERIAL COST							
TOTAL LABOR COST - GROUP 1	0.60 WH		Hr				
GROUP 2	0.90 WH		Hr				
GROUP 3	1.20 WH		Hr				
GROUP 4	1.50 WH		Hr				
DIRECT JOB EXPENSE							
TOTAL PRIME COST				$	$	$	$

Receptacle – Floor Mounted

Cost Item	Quantity	Price or Rate	Per	Installation Groups 1	2	3	4
MATERIAL							
Box and support	1		E				
#14-2 w/grd. Type NM cable	20 - 30 ft.		C Ft.				
Floor receptacle and cover	1		E				
Miscellaneous	Lot						
TOTAL MATERIAL COST							
TOTAL LABOR COST - GROUP 1	1.15 WH		Hr				
GROUP 2	1.65 WH		Hr				
GROUP 3	2.25 WH		Hr				
GROUP 4	2.75 WH		Hr				
DIRECT JOB EXPENSE							
TOTAL PRIME COST				$	$	$	$

Receptacle – Floor Mounted

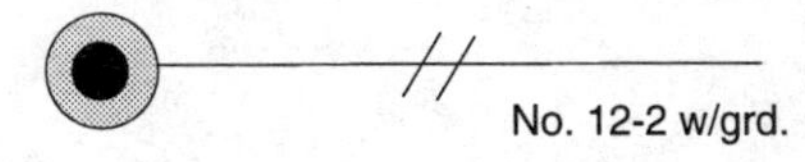

Cost Item	Quantity	Price or Rate	Per	Installation Groups			
MATERIAL				1	2	3	4
Box and support	1		E				
#12-2 w/grd. Type NM cable	20 - 30 ft.		C Ft.				
Floor receptacle and cover	1		E				
Miscellaneous	Lot						
TOTAL MATERIAL COST							
TOTAL LABOR COST - GROUP 1	1.15 WH		Hr				
GROUP 2	1.70 WH		Hr				
GROUP 3	2.15 WH		Hr				
GROUP 4	2.80 WH		Hr				
DIRECT JOB EXPENSE							
TOTAL PRIME COST				$	$	$	$

Duplex Receptacle – Clock Hanger

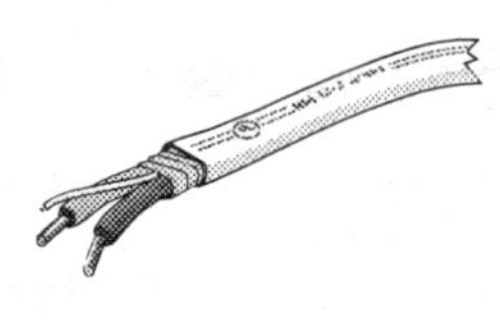

Cost Item	Quantity	Price or Rate	Per	Installation Groups 1	2	3	4
MATERIAL							
Box and support	1		E				
#14-2 w/grd. Type NM cable	20 - 30 ft.		C Ft.				
Clock hanger receptacle and cover	1		E				
Miscellaneous	Lot						
TOTAL MATERIAL COST							
TOTAL LABOR COST - GROUP 1	0.55 WH		Hr				
GROUP 2	0.85 WH		Hr				
GROUP 3	1.15 WH		Hr				
GROUP 4	1.45 WH		Hr				
DIRECT JOB EXPENSE							
TOTAL PRIME COST				$	$	$	$

Duplex Receptacle – Clock Hanger

Cost Item	Quantity	Price or Rate	Per	Installation Groups 1	2	3	4
MATERIAL							
Box and support	1		E				
#12-2 w/grd. Type NM cable	20 - 30 ft.		C Ft.				
Clock hanger receptacle and cover	1		E				
Miscellaneous	Lot						
TOTAL MATERIAL COST							
TOTAL LABOR COST - GROUP 1	0.65 WH		Hr				
GROUP 2	0.95 WH		Hr				
GROUP 3	1.25 WH		Hr				
GROUP 4	1.55 WH		Hr				
DIRECT JOB EXPENSE							
TOTAL PRIME COST				$	$	$	$

Ceiling Fan Receptacle

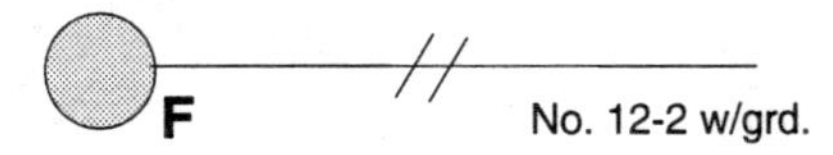

Cost Item	Quantity	Price or Rate	Per	Installation Groups 1	2	3	4
MATERIAL							
Ceiling fan box and support	1		E				
#12-2 w/grd. Type NM cable	20 - 30 ft.		C Ft.				
Ceiling fan receptacle and cover	1		E				
Miscellaneous	Lot						
TOTAL MATERIAL COST							
TOTAL LABOR COST - GROUP 1	1.20 WH		Hr				
GROUP 2	1.85 WH		Hr				
GROUP 3	2.40 WH		Hr				
GROUP 4	3.00 WH		Hr				
DIRECT JOB EXPENSE							
TOTAL PRIME COST				$	$	$	$

Wall Switch – Single-Pole

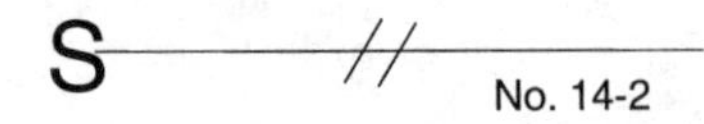

Cost Item	Quantity	Price or Rate	Per	Installation Groups 1	2	3	4
MATERIAL							
Box and support	1		E				
#14-2 w/grd. Type NM cable	20 - 30 ft.		C Ft.				
Single-pole switch and plate	1		E				
Miscellaneous	Lot						
TOTAL MATERIAL COST							
TOTAL LABOR COST - GROUP 1	0.45 WH		Hr				
GROUP 2	0.70 WH		Hr				
GROUP 3	0.90 WH		Hr				
GROUP 4	1.15 WH		Hr				
DIRECT JOB EXPENSE							
TOTAL PRIME COST				$	$	$	$

Wall Switch – Single-Pole

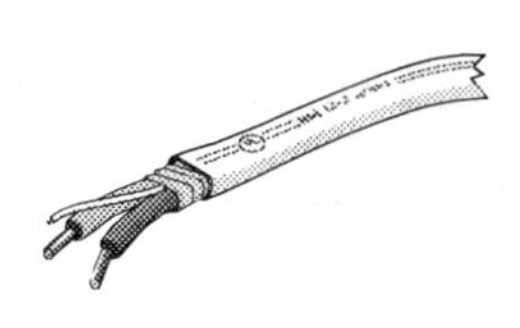

Cost Item	Quantity	Price or Rate	Per	Installation Groups 1	2	3	4
MATERIAL							
Box and support	1		E				
#12-2 w/grd. Type NM cable	20 - 30 ft.		C Ft.				
Single-pole switch with plate	1		E				
Miscellaneous	Lot						
TOTAL MATERIAL COST							
TOTAL LABOR COST - GROUP 1	0.55 WH		Hr				
GROUP 2	0.80 WH		Hr				
GROUP 3	1.00 WH		Hr				
GROUP 4	1.30 WH		Hr				
DIRECT JOB EXPENSE							
TOTAL PRIME COST				$	$	$	$

Wall Switch with Pilot Light

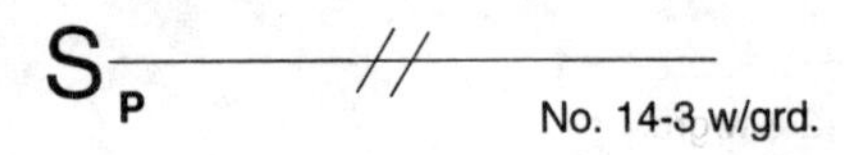

Cost Item	Quantity	Price or Rate	Per	Installation Groups 1	2	3	4
MATERIAL							
Box and support	1		E				
#14-3 w/grd. Type NM cable	20 - 30 ft.		C Ft.				
Combination switch/pilot light with plate	1		E				
Miscellaneous	Lot						
TOTAL MATERIAL COST							
TOTAL LABOR COST - GROUP 1	0.55 WH		Hr				
GROUP 2	0.85 WH		Hr				
GROUP 3	1.15 WH		Hr				
GROUP 4	1.45 WH		Hr				
DIRECT JOB EXPENSE							
TOTAL PRIME COST				$	$	$	$

Wall Switch with Pilot Light

Cost Item	Quantity	Price or Rate	Per	Installation Groups 1	2	3	4
MATERIAL							
Box and support	1		E				
#12-3 w/grd. Type NM cable	20 - 30 ft.		C Ft.				
Combination switch/pilot light with plate	1						
Miscellaneous	Lot						
TOTAL MATERIAL COST							
TOTAL LABOR COST - GROUP 1	0.65 WH		Hr				
GROUP 2	0.95 WH		Hr				
GROUP 3	1.25 WH		Hr				
GROUP 4	1.50 WH		Hr				
DIRECT JOB EXPENSE							
TOTAL PRIME COST				$	$	$	$

Wall Switch – Three-Way

S_3 ——///—— No. 14-3 w/grd.

Cost Item	Quantity	Price or Rate	Per	Installation Groups			
				1	2	3	4
MATERIAL							
Box and support	1		E				
#14-3 w/grd. Type NM cable	25 - 35 ft.		C Ft.				
Three-way switch with plate	1		E				
Miscellaneous	Lot						
TOTAL MATERIAL COST							
TOTAL LABOR COST - GROUP 1	0.65 WH		Hr				
GROUP 2	1.00 WH		Hr				
GROUP 3	1.35 WH		Hr				
GROUP 4	1.70 WH		Hr				
DIRECT JOB EXPENSE							
TOTAL PRIME COST				$	$	$	$

Wall Switch – Three-Way

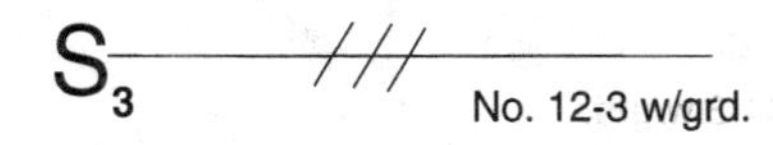

Cost Item	Quantity	Price or Rate	Per	Installation Groups 1	2	3	4
MATERIAL							
Box and support	1		E				
#12-3 w/grd. Type NM cable	25 - 35 ft.		C Ft.				
Three-way switch with plate	1		E				
Miscellaneous	Lot						
TOTAL MATERIAL COST							
TOTAL LABOR COST - GROUP 1	0.75 WH		Hr				
GROUP 2	1.10 WH		Hr				
GROUP 3	1.45 WH		Hr				
GROUP 4	1.80 WH		Hr				
DIRECT JOB EXPENSE							
TOTAL PRIME COST				$	$	$	$

Wall Switch – Four-Way

S_4 ——///—— No. 14-3 w/grd.

Cost Item	Quantity	Price or Rate	Per	Installation Groups 1	2	3	4
MATERIAL							
Box and support	1		E				
#14-3 w/grd. Type NM cable	35 - 50 ft.		C Ft.				
Four-way switch and plate	1						
Miscellaneous	Lot						
TOTAL MATERIAL COST							
TOTAL LABOR COST - GROUP 1	0.65 WH		Hr				
GROUP 2	1.00 WH		Hr				
GROUP 3	1.35 WH		Hr				
GROUP 4	1.70 WH		Hr				
DIRECT JOB EXPENSE							
TOTAL PRIME COST				$	$	$	$

Wall Switch – Four-Way

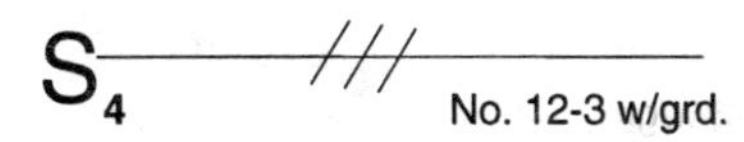

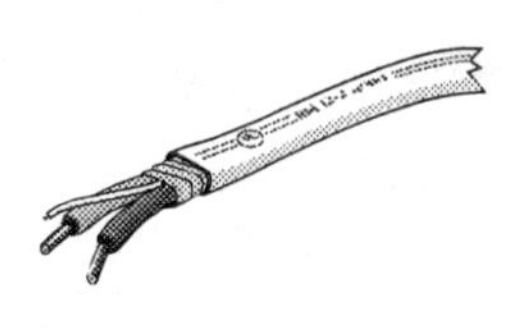

Cost Item	Quantity	Price or Rate	Per	Installation Groups			
				1	2	3	4
MATERIAL							
Box and support	1		E				
#12-3 w/grd. Type NM cable	35 - 50 ft.		C Ft.				
Four-way swith with plate	1		E				
Miscellaneous	Lot						
TOTAL MATERIAL COST							
TOTAL LABOR COST - GROUP 1	0.75 WH		Hr				
GROUP 2	1.10 WH		Hr				
GROUP 3	1.50 WH		Hr				
GROUP 4	1.75 WH		Hr				
DIRECT JOB EXPENSE							
TOTAL PRIME COST				$	$	$	$

Dimmer Switch/Control

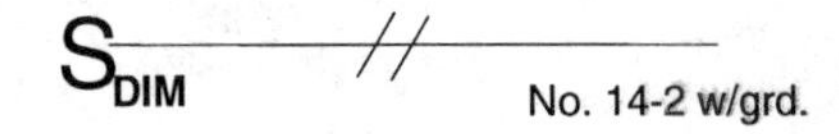

Cost Item	Quantity	Price or Rate	Per	Installation Groups			
				1	2	3	4
MATERIAL							
Box and support	1		E				
#14-2 w/grd. Type NM cable	20 - 35 ft.		C Ft.				
Dimmer control and cover	1		E				
Miscellaneous	Lot						
TOTAL MATERIAL COST							
TOTAL LABOR COST - GROUP 1	0.85 WH		Hr				
GROUP 2	1.25 WH		Hr				
GROUP 3	1.75 WH		Hr				
GROUP 4	2.20 WH		Hr				
DIRECT JOB EXPENSE							
TOTAL PRIME COST				$	$	$	$

Dimmer Switch/Control

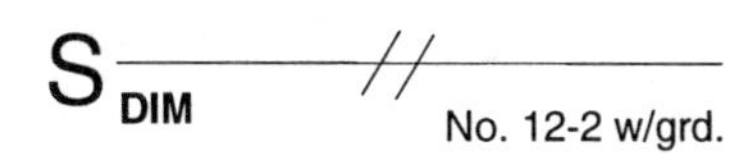

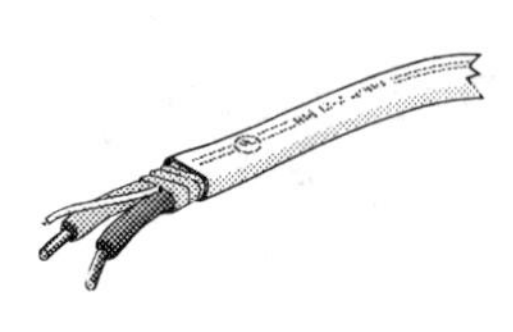

Cost Item	Quantity	Price or Rate	Per	Installation Groups 1	2	3	4
MATERIAL							
Box and support	1		E				
#12-2 w/grd. Type NM cable	15 - 35 ft.		C Ft.				
Dimmer control with cover	1		E				
Miscellaneous	Lot						
TOTAL MATERIAL COST							
TOTAL LABOR COST - GROUP 1	0.95 WH		Hr				
GROUP 2	1.25 WH		Hr				
GROUP 3	1.85 WH		Hr				
GROUP 4	2.30 WH		Hr				
DIRECT JOB EXPENSE							
TOTAL PRIME COST				$	$	$	$

Door Switch

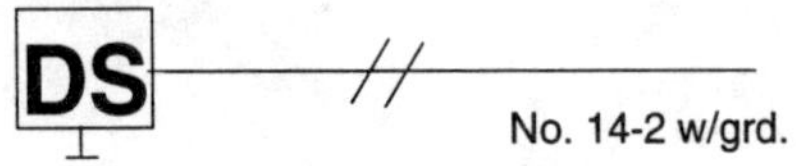

Cost Item	Quantity	Price or Rate	Per	Installation Groups 1	2	3	4
MATERIAL							
Box and support	1		E				
#14-2 w/grd. Type NM cable	15 - 35 ft.		C Ft.				
Door switch complete with box, cover and striking plate	1		E				
Miscellaneous	Lot						
TOTAL MATERIAL COST							
TOTAL LABOR COST - GROUP 1	0.85 WH		Hr				
GROUP 2	1.30 WH		Hr				
GROUP 3	1.75 WH		Hr				
GROUP 4	2.20 WH		Hr				
DIRECT JOB EXPENSE							
TOTAL PRIME COST				$	$	$	$

Door Switch

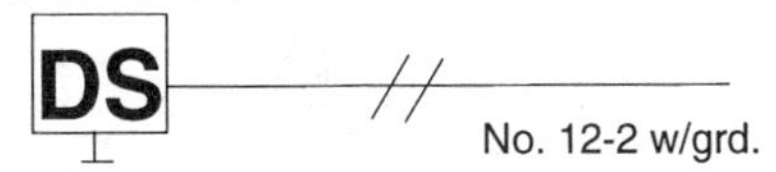

Cost Item	Quantity	Price or Rate	Per	Installation Groups 1	2	3	4
MATERIAL							
Box and support	1		E				
#12-2 w/grd. Type NM cable	20 - 35 ft.		C Ft.				
Door switch complete with box, cover and striking plate	1		E				
Miscellaneous	Lot						
TOTAL MATERIAL COST							
TOTAL LABOR COST - GROUP 1	0.95 WH		Hr				
GROUP 2	1.40 WH		Hr				
GROUP 3	1.85 WH		Hr				
GROUP 4	2.30 WH		Hr				
DIRECT JOB EXPENSE							
TOTAL PRIME COST				$	$	$	$

Weatherproof Switch

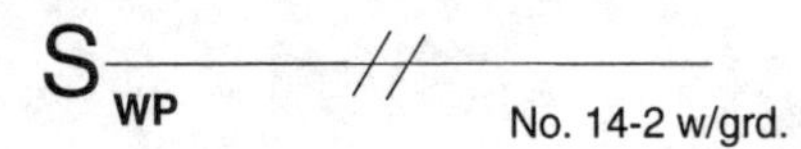

Cost Item	Quantity	Price or Rate	Per	Installation Groups 1	2	3	4
MATERIAL							
Box and support	1		E				
#14-2 w/grd. Type NM cable	35 - 50 ft.		C Ft.				
Switch and weatherproof plate	1		E				
Miscellaneous	Lot						
TOTAL MATERIAL COST							
TOTAL LABOR COST - GROUP 1	0.65 WH		Hr				
GROUP 2	1.00 WH		Hr				
GROUP 3	1.35 WH		Hr				
GROUP 4	1.70 WH		Hr				
DIRECT JOB EXPENSE							
TOTAL PRIME COST				$	$	$	$

Weatherproof Switch

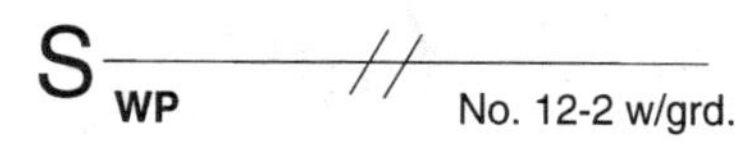

Cost Item	Quantity	Price or Rate	Per	Installation Groups 1	2	3	4
MATERIAL							
Box and support	1		E				
#12-2 w/grd. Type NM cable	20 - 35 ft.		C Ft.				
Switch with weatherproof cover	1		E				
Miscellaneous	Lot						
TOTAL MATERIAL COST							
TOTAL LABOR COST - GROUP 1	0.75 WH		Hr				
GROUP 2	1.10 WH		Hr				
GROUP 3	1.45 WH		Hr				
GROUP 4	1.75 WH		Hr				
DIRECT JOB EXPENSE							
TOTAL PRIME COST				$	$	$	$

Branch Circuit to Fixed Equipment

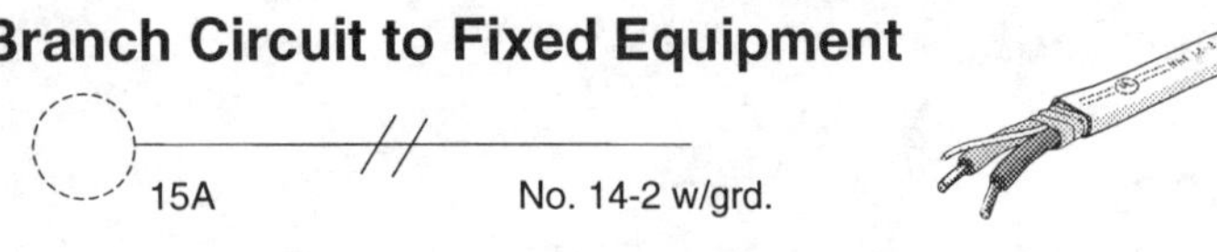

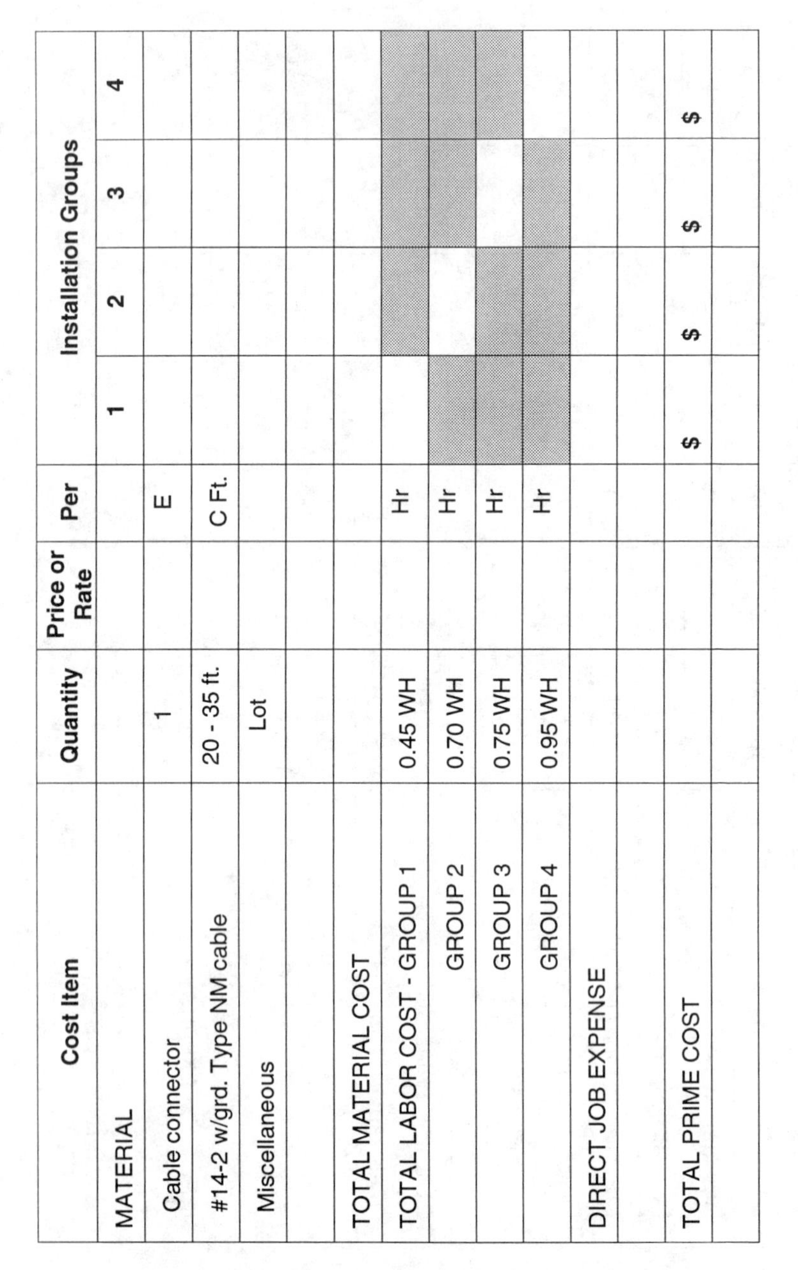

Cost Item	Quantity	Price or Rate	Per	Installation Groups 1	2	3	4
MATERIAL							
Cable connector	1		E				
#14-2 w/grd. Type NM cable	20 - 35 ft.		C Ft.				
Miscellaneous	Lot						
TOTAL MATERIAL COST							
TOTAL LABOR COST - GROUP 1	0.45 WH		Hr				
GROUP 2	0.70 WH		Hr				
GROUP 3	0.75 WH		Hr				
GROUP 4	0.95 WH		Hr				
DIRECT JOB EXPENSE							
TOTAL PRIME COST				$	$	$	$

Branch Circuit to Fixed Equipment

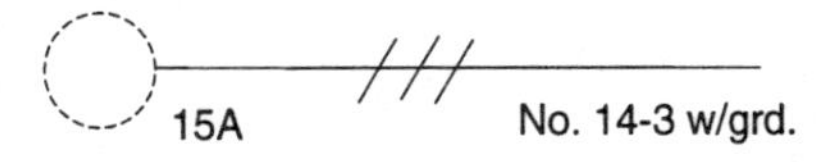

Cost Item	Quantity	Price or Rate	Per	Installation Groups 1	2	3	4
MATERIAL							
Cable connector	1		E				
#14-3 w/grd. Type NM cable	20 - 35 ft.		C Ft.				
Miscellaneous	Lot						
TOTAL MATERIAL COST							
TOTAL LABOR COST - GROUP 1	0.55 WH		Hr				
GROUP 2	0.80 WH		Hr				
GROUP 3	1.05 WH		Hr				
GROUP 4	1.30 WH		Hr				
DIRECT JOB EXPENSE							
TOTAL PRIME COST				$	$	$	$

Branch Circuit to Fixed Equipment

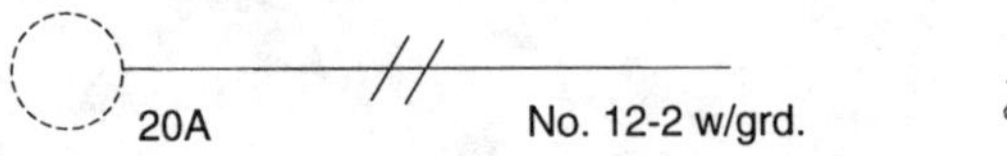

Cost Item	Quantity	Price or Rate	Per	Installation Groups 1	2	3	4
MATERIAL							
Cable connector	1		E				
#12-2 w/grd. Type NM cable	20 - 35 ft.		C Ft.				
Miscellaneous	Lot						
TOTAL MATERIAL COST							
TOTAL LABOR COST - GROUP 1	0.45 WH		Hr				
GROUP 2	0.65 WH		Hr				
GROUP 3	0.85 WH		Hr				
GROUP 4	1.10 WH		Hr				
DIRECT JOB EXPENSE							
TOTAL PRIME COST				$	$	$	$

Branch Circuit to Fixed Equipment

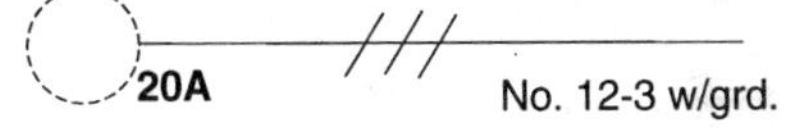

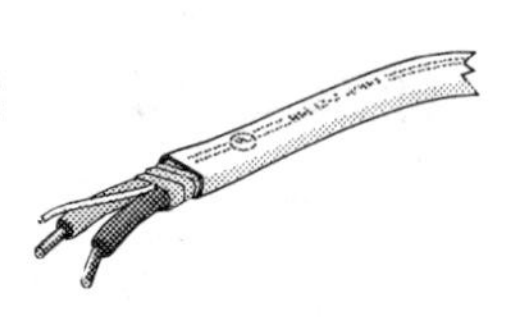

Cost Item	Quantity	Price or Rate	Per	Installation Groups			
				1	2	3	4
MATERIAL							
Cable connector	1		E				
#12-3 w/grd. Type NM Cable	20 - 35 ft.		C Ft.				
Miscellaneous	Lot						
TOTAL MATERIAL COST							
TOTAL LABOR COST - GROUP 1	0.55 WH		Hr				
GROUP 2	0.80 WH		Hr				
GROUP 3	1.05 WH		Hr				
GROUP 4	1.30 WH		Hr				
DIRECT JOB EXPENSE							
TOTAL PRIME COST				$	$	$	$

Branch Circuit to Fixed Equipment

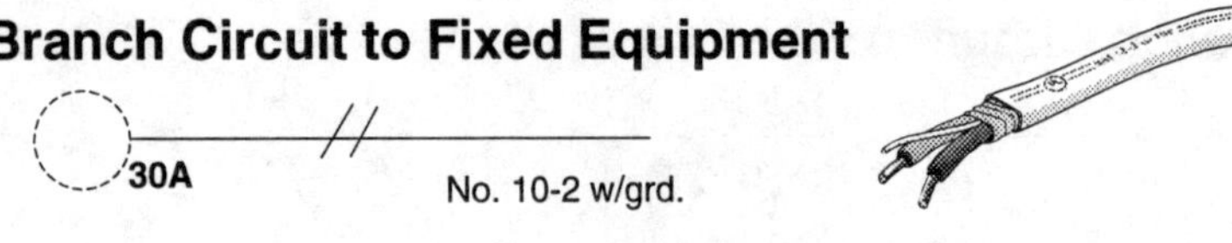

Cost Item	Quantity	Price or Rate	Per	Installation Groups			
				1	2	3	4
MATERIAL							
Cable connector	1		E				
#10-2 w/grd. Type NM Cable	20 - 35 ft.		C Ft.				
Miscellaneous	Lot						
TOTAL MATERIAL COST							
TOTAL LABOR COST - GROUP 1	0.50 WH		Hr				
GROUP 2	0.70 WH		Hr				
GROUP 3	0.90 WH		Hr				
GROUP 4	1.10 WH		Hr				
DIRECT JOB EXPENSE							
TOTAL PRIME COST				$	$	$	$

Branch Circuit to Fixed Equipment

Cost Item	Quantity	Price or Rate	Per	Installation Groups 1	2	3	4
MATERIAL							
Cable connector	1		E				
#10-3 w/grd. Type NM Cable	20 - 35 ft.		C Ft.				
Miscellaneous	Lot						
TOTAL MATERIAL COST							
TOTAL LABOR COST - GROUP 1	0.60 WH		Hr				
GROUP 2	0.80 WH		Hr				
GROUP 3	1.00 WH		Hr				
GROUP 4	1.20 WH		Hr				
DIRECT JOB EXPENSE							
TOTAL PRIME COST				$	$	$	$

Branch Circuit to Special Receptacle

20A

No. 12-3 w/grd.

Cost Item	Quantity	Price or Rate	Per	Installation Groups 1	2	3	4
MATERIAL							
Box and support	1		E				
#12-3 w/grd. Type NM Cable	20 - 35 ft.		C Ft.				
20-amp special-purpose receptable	1		E				
Miscellaneous	Lot						
TOTAL MATERIAL COST							
TOTAL LABOR COST - GROUP 1	1.00 WH		Hr				
GROUP 2	1.50 WH		Hr				
GROUP 3	2.00 WH		Hr				
GROUP 4	2.50 WH		Hr				
DIRECT JOB EXPENSE							
TOTAL PRIME COST				$	$	$	$

Branch Circuit to Special Receptacle

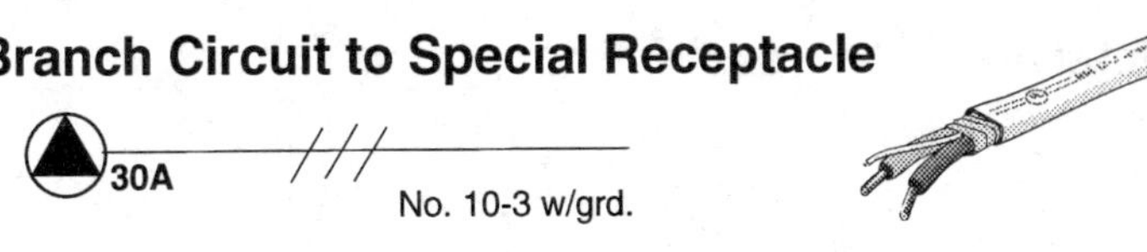

Cost Item	Quantity	Price or Rate	Per	Installation Groups 1	2	3	4
MATERIAL							
Box and support	1		E				
#10-3 w/grd. Type NM Cable	20 - 35 ft.		C Ft.				
30-amp special-purpose receptacle							
Miscellaneous	Lot						
TOTAL MATERIAL COST							
TOTAL LABOR COST - GROUP 1	2.00 WH		Hr				
GROUP 2	3.00 WH		Hr				
GROUP 3	4.00 WH		Hr				
GROUP 4	5.00 WH		Hr				
DIRECT JOB EXPENSE							
TOTAL PRIME COST				$	$	$	$

Branch Circuit to Special Receptacle

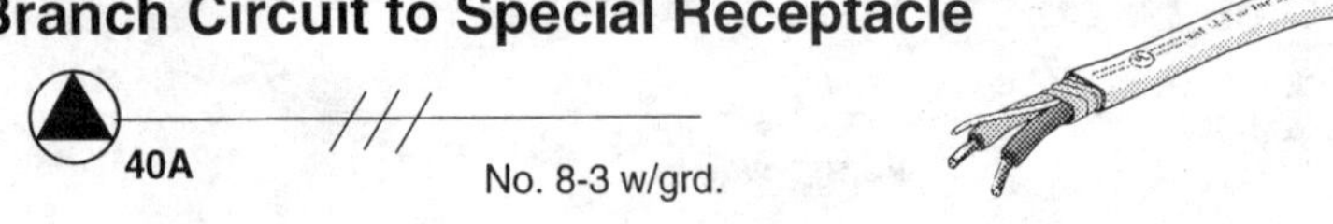

Cost Item	Quantity	Price or Rate	Per	Installation Groups 1	2	3	4
MATERIAL							
Box and support	1		E				
#8-3 w/grd. Type NM Cable	20 - 35 ft.		C Ft.				
40-amp special-purpose receptacle	1		E				
Miscellaneous	Lot						
TOTAL MATERIAL COST							
TOTAL LABOR COST - GROUP 1	2.55 WH		Hr				
GROUP 2	3.80 WH		Hr				
GROUP 3	4.05 WH		Hr				
GROUP 4	5.30 WH		Hr				
DIRECT JOB EXPENSE							
TOTAL PRIME COST				$	$	$	$

Branch Circuit to Range Receptacle

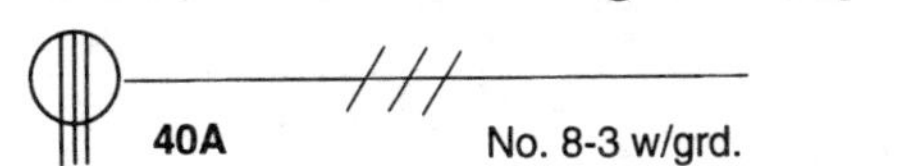

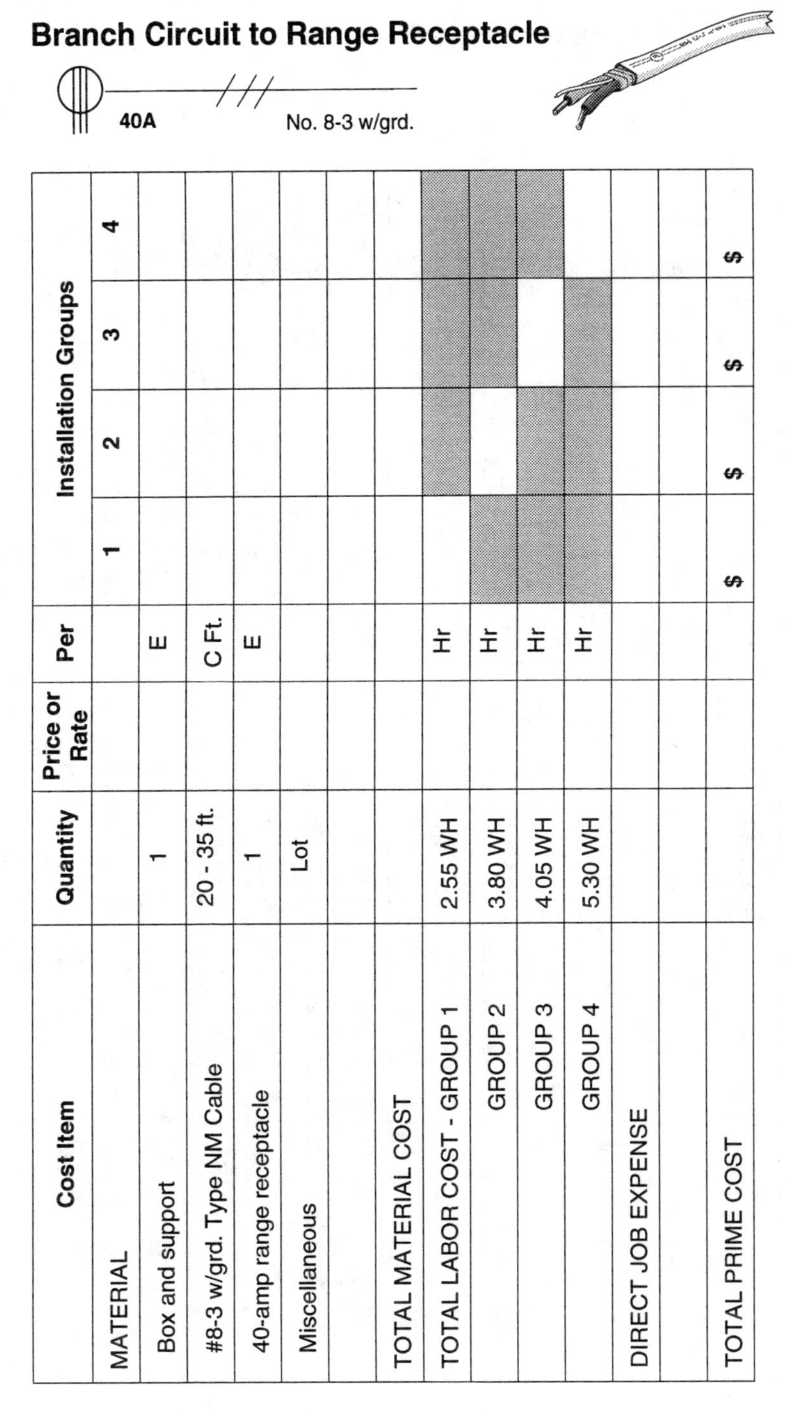

Cost Item	Quantity	Price or Rate	Per	Installation Groups			
				1	2	3	4
MATERIAL							
Box and support	1		E				
#8-3 w/grd. Type NM Cable	20 - 35 ft.		C Ft.				
40-amp range receptacle	1		E				
Miscellaneous	Lot						
TOTAL MATERIAL COST							
TOTAL LABOR COST - GROUP 1	2.55 WH		Hr				
GROUP 2	3.80 WH		Hr				
GROUP 3	4.05 WH		Hr				
GROUP 4	5.30 WH		Hr				
DIRECT JOB EXPENSE							
TOTAL PRIME COST				$	$	$	$

Branch Circuit to Range Receptacle

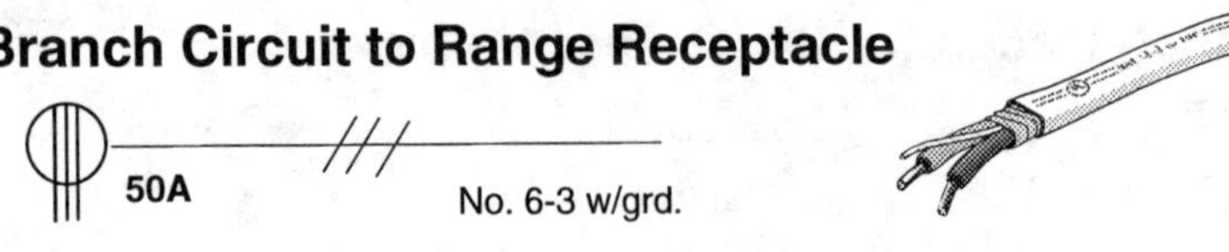

Cost Item	Quantity	Price or Rate	Per	Installation Groups 1	2	3	4
MATERIAL							
Box and support	1		E				
#6-3 w/grd. Type NM Cable	20 - 35 ft.		C Ft.				
50-amp range receptacle	1		E				
Miscellaneous	Lot						
TOTAL MATERIAL COST							
TOTAL LABOR COST - GROUP 1	3.55 WH		Hr				
GROUP 2	5.30 WH		Hr				
GROUP 3	7.05 WH		Hr				
GROUP 4	8.30 WH		Hr				
DIRECT JOB EXPENSE							
TOTAL PRIME COST				$	$	$	$

Feeder – 240V, 3-Wire, Single-Phase, 40-Amp

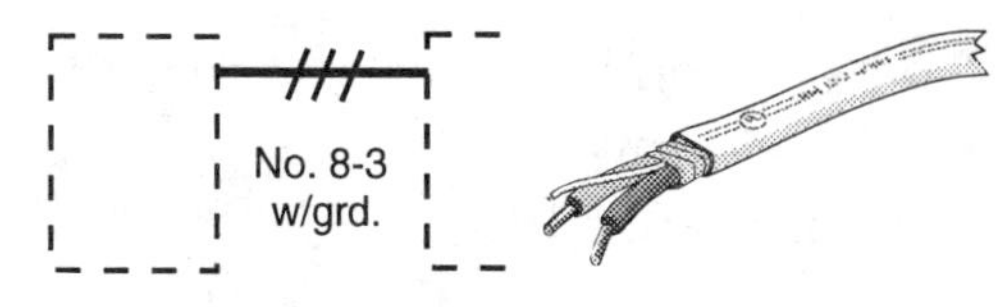

Cost Item	Quantity	Price or Rate	Per	Installation Groups 1	2	3	4
MATERIAL							
Cable connectors	2		E				
#8-3 w/grd. Type NM Cable	20 - 35 ft.		C Ft.				
Miscellaneous	Lot						
TOTAL MATERIAL COST							
TOTAL LABOR COST - GROUP 1	2.05 WH		Hr				
GROUP 2	3.00 WH		Hr				
GROUP 3	4.05 WH		Hr				
GROUP 4	5.10 WH		Hr				
DIRECT JOB EXPENSE							
TOTAL PRIME COST				$	$	$	$

Feeder – 240V, 3-Wire, Single-Phase, 60-Amp

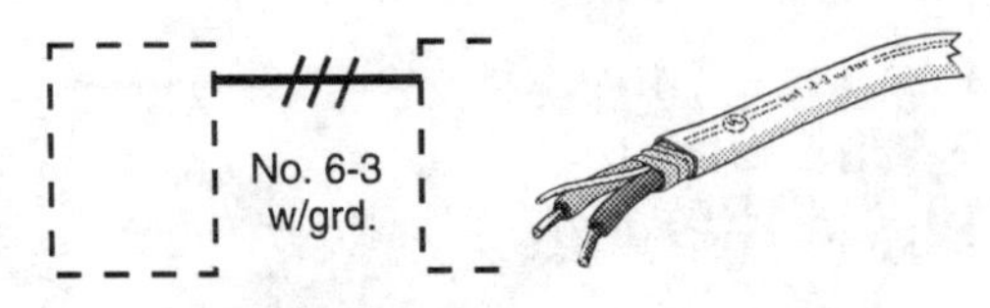

Cost Item	Quantity	Price or Rate	Per	Installation Groups			
				1	2	3	4
MATERIAL							
Cable connectors	2		E				
#6-3 w/grd. Type NM Cable	20 - 35 ft.		C Ft.				
Miscellaneous	Lot						
TOTAL MATERIAL COST							
TOTAL LABOR COST - GROUP 1	2.30 WH		Hr				
GROUP 2	3.50 WH		Hr				
GROUP 3	4.50 WH		Hr				
GROUP 4	5.70 WH		Hr				
DIRECT JOB EXPENSE							
TOTAL PRIME COST				$	$	$	$

Feeder – 240V, 3-Wire, Single-Phase, 100-Amp

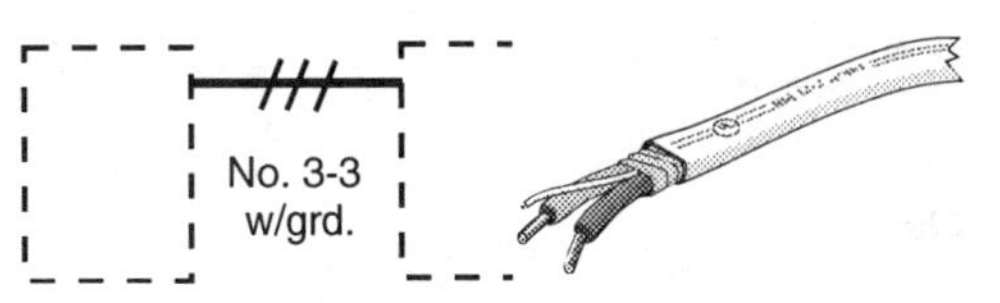

Cost Item	Quantity	Price or Rate	Per	Installation Groups 1	2	3	4
MATERIAL							
Cable connectors	2		E				
#3-3 w/grd. Type NM Cable	20 - 35 ft.		C Ft.				
Miscellaneous	Lot						
TOTAL MATERIAL COST							
TOTAL LABOR COST - GROUP 1	3.00 WH		Hr				
GROUP 2	4.80 WH		Hr				
GROUP 3	5.05 WH		Hr				
GROUP 4	6.30 WH		Hr				
DIRECT JOB EXPENSE							
TOTAL PRIME COST				$	$	$	$

Feeder – 240V, 3-Wire, Single-Phase, 125-Amp

Cost Item	Quantity	Price or Rate	Per	Installation Groups			
				1	2	3	4
MATERIAL							
Cable connectors	2		E				
#1-3 w/grd. Type NM Cable	20 - 35 ft.		C Ft.				
Miscellaneous	Lot						
TOTAL MATERIAL COST							
TOTAL LABOR COST - GROUP 1	3.55 WH		Hr				
GROUP 2	5.00 WH		Hr				
GROUP 3	5.50 WH		Hr				
GROUP 4	6.50 WH		Hr				
DIRECT JOB EXPENSE							
TOTAL PRIME COST				$	$	$	$

Selling-Price Tables

The selling-price tables to follow are designed for quick unit pricing of the various outlets found in this chapter, using nonmetallic-sheathed cable. Before arriving at a selling price, the preceding tables must be completed (filled in with appropriate prices). Then the prime cost determined for each wiring situation in the preceding tables should be entered in the corresponding selling-price tables to follow. Once the prime cost has been entered in each selling-price table, overhead and profit factors are calculated to arrive at a total unit selling price as described in detail in Chapter 2.

Surface-Mounted Lighting Outlet	#14-2	GROUPS			
		1	2	3	4
Total Prime Cost		$	$	$	$
Overhead	_% of prime cost				
Total Cost		$	$	$	$
Profit	_% of total cost				
Selling Price		$	$	$	$

Surface-Mounted Lighting Outlet	#12-2	GROUPS			
		1	2	3	4
Total Prime Cost		$	$	$	$
Overhead	_% of prime cost				
Total Cost		$	$	$	$
Profit	_% of total cost				
Selling Price		$	$	$	$

Recessed Lighting Outlet, 15-Amp	#14-2	GROUPS			
		1	2	3	4
Total Prime Cost		$	$	$	$
Overhead	_% of prime cost				
Total Cost		$	$	$	$
Profit	_% of total cost				
Selling Price		$	$	$	$

Recessed Lighting Outlet, 20-Amp	#12-2	GROUPS			
		1	2	3	4
Total Prime Cost		$	$	$	$
Overhead	_% of prime cost				
Total Cost		$	$	$	$
Profit	_% of total cost				
Selling Price		$	$	$	$

Wall-Mounted Lighting Outlet, 15-Amp	#14-2	GROUPS			
		1	2	3	4
Total Prime Cost		$	$	$	$
Overhead	_% of prime cost				
Total Cost		$	$	$	$
Profit	_% of total cost				
Selling Price		$	$	$	$

Wall-Mounted Lighting Outlet, 20-Amp	#12-2	GROUPS			
		1	2	3	4
Total Prime Cost		$	$	$	$
Overhead	_% of prime cost				
Total Cost		$	$	$	$
Profit	_% of total cost				
Selling Price		$	$	$	$

Duplex Receptacle 2-Wire Circuit, 15-Amp	#14-2	GROUPS			
		1	2	3	4
Total Prime Cost		$	$	$	$
Overhead	_% of prime cost				
Total Cost		$	$	$	$
Profit	_% of total cost				
Selling Price		$	$	$	$

Duplex Receptacle 2-Wire Circuit, 20-Amp	#12-2	GROUPS			
		1	2	3	4
Total Prime Cost		$	$	$	$
Overhead	_% of prime cost				
Total Cost		$	$	$	$
Profit	_% of total cost				
Selling Price		$	$	$	$

Duplex Receptacle 3-Wire Circuit, 15-Amp	#14-3	GROUPS			
		1	2	3	4
Total Prime Cost		$	$	$	$
Overhead	_% of prime cost				
Total Cost		$	$	$	$
Profit	_% of total cost				
Selling Price		$	$	$	$

Duplex Receptacle 3-Wire Circuit, 20-Amp	#12-3	**GROUPS**			
		1	**2**	**3**	**4**
Total Prime Cost		$	$	$	$
Overhead	_% of prime cost				
Total Cost		$	$	$	$
Profit	_% of total cost				
Selling Price		$	$	$	$

Duplex Receptacle Split-Wired, 15-Amp	#14-3	**GROUPS**			
		1	**2**	**3**	**4**
Total Prime Cost		$	$	$	$
Overhead	_% of prime cost				
Total Cost		$	$	$	$
Profit	_% of total cost				
Selling Price		$	$	$	$

Duplex Receptacle Split-Wired, 20-Amp	#12-3	**GROUPS**			
		1	**2**	**3**	**4**
Total Prime Cost		$	$	$	$
Overhead	_% of prime cost				
Total Cost		$	$	$	$
Profit	_% of total cost				
Selling Price		$	$	$	$

Duplex Receptacle Weatherproof 15-Amp	#14-2	**GROUPS**			
		1	**2**	**3**	**4**
Total Prime Cost		$	$	$	$
Overhead	_% of prime cost				
Total Cost		$	$	$	$
Profit	_% of total cost				
Selling Price		$	$	$	$

Duplex Receptacle Weatherproof, 20-Amp	#12-2	**GROUPS**			
		1	**2**	**3**	**4**
Total Prime Cost		$	$	$	$
Overhead	__ % of prime cost				
Total Cost		$	$	$	$
Profit	__ % of total cost				
Selling Price		$	$	$	$

Duplex Receptacle GFCI, 15-Amp	#14-2	**GROUPS**			
		1	**2**	**3**	**4**
Total Prime Cost		$	$	$	$
Overhead	__% of prime cost				
Total Cost		$	$	$	$
Profit	__% of total cost				
Selling Price		$	$	$	$

Duplex Receptacle GFCI, 20-Amp	#12-2	**GROUPS**			
		1	**2**	**3**	**4**
Total Prime Cost		**$**	**$**	**$**	**$**
Overhead	__ % of prime cost				
Total Cost		**$**	**$**	**$**	**$**
Profit	__ % of total cost				
Selling Price		**$**	**$**	**$**	**$**

Duplex Receptacle Floor Mounted, 15-Amp	#14-2	**GROUPS**			
		1	**2**	**3**	**4**
Total Prime Cost		**$**	**$**	**$**	**$**
Overhead	__ % of prime cost				
Total Cost		**$**	**$**	**$**	**$**
Profit	__ % of total cost				
Selling Price		**$**	**$**	**$**	**$**

Duplex Receptacle Floor Mounted, 20-Amp	#12-2	**GROUPS**			
		1	**2**	**3**	**4**
Total Prime Cost		**$**	**$**	**$**	**$**
Overhead	__ % of prime cost				
Total Cost		**$**	**$**	**$**	**$**
Profit	__ % of total cost				
Selling Price		**$**	**$**	**$**	**$**

Duplex Receptacle Clock Hanger, 15-Amp	#14-2	**GROUPS**			
		1	**2**	**3**	**4**
Total Prime Cost		**$**	**$**	**$**	**$**
Overhead	_% of prime cost				
Total Cost		**$**	**$**	**$**	**$**
Profit	_% of total cost				
Selling Price		**$**	**$**	**$**	**$**

Duplex Receptacle Clock Hanger, 20-Amp	#12-2	**GROUPS**			
		1	**2**	**3**	**4**
Total Prime Cost		**$**	**$**	**$**	**$**
Overhead	_% of prime cost				
Total Cost		**$**	**$**	**$**	**$**
Profit	_% of total cost				
Selling Price		**$**	**$**	**$**	**$**

Duplex Receptacle Fan Receptacle, 20-Amp	#12-2	**GROUPS**			
		1	**2**	**3**	**4**
Total Prime Cost		**$**	**$**	**$**	**$**
Overhead	_% of prime cost				
Total Cost		**$**	**$**	**$**	**$**
Profit	_% of total cost				
Selling Price		**$**	**$**	**$**	**$**

Wall Switch Single-Pole, 15-Amp	#14-2	GROUPS			
		1	2	3	4
Total Prime Cost		$	$	$	$
Overhead	_% of prime cost				
Total Cost		$	$	$	$
Profit	_% of total cost				
Selling Price		$	$	$	$

Wall Switch Single-Pole, 20-Amp	#12-2	GROUPS			
		1	2	3	4
Total Prime Cost		$	$	$	$
Overhead	_% of prime cost				
Total Cost		$	$	$	$
Profit	_% of total cost				
Selling Price		$	$	$	$

Wall Switch with Pilot Light, 15-Amp	#14-3	GROUPS			
		1	2	3	4
Total Prime Cost		$	$	$	$
Overhead	_% of prime cost				
Total Cost		$	$	$	$
Profit	_% of total cost				
Selling Price		$	$	$	$

Wall Switch with Pilot Light, 20-Amp	#12-3	GROUPS			
		1	2	3	4
Total Prime Cost		$	$	$	$
Overhead	_% of prime cost				
Total Cost		$	$	$	$
Profit	_% of total cost				
Selling Price		$	$	$	$

Wall Switch Three-Way, 15-Amp	#14-3	GROUPS			
		1	2	3	4
Total Prime Cost		$	$	$	$
Overhead	_% of prime cost				
Total Cost		$	$	$	$
Profit	_% of total cost				
Selling Price		$	$	$	$

Wall Switch Three-Way, 20-Amp	#12-3	GROUPS			
		1	2	3	4
Total Prime Cost		$	$	$	$
Overhead	_% of prime cost				
Total Cost		$	$	$	$
Profit	_% of total cost				
Selling Price		$	$	$	$

Wall Switch Four-Way, 15-Amp	#14-3	**GROUPS**			
		1	**2**	**3**	**4**
Total Prime Cost		$	$	$	$
Overhead	_% of prime cost				
Total Cost		$	$	$	$
Profit	_% of total cost				
Selling Price		$	$	$	$

Wall Switch Four-Way, 20-Amp	#12-3	**GROUPS**			
		1	**2**	**3**	**4**
Total Prime Cost		$	$	$	$
Overhead	_% of prime cost				
Total Cost		$	$	$	$
Profit	_% of total cost				
Selling Price		$	$	$	$

Dimmer Switch/Control, 15-Amp	#14-2	**GROUPS**			
		1	**2**	**3**	**4**
Total Prime Cost		$	$	$	$
Overhead	_% of prime cost				
Total Cost		$	$	$	$
Profit	_% of total cost				
Selling Price		$	$	$	$

Dimmer Switch/Control, 20-Amp	#12-2	GROUPS			
		1	2	3	4
Total Prime Cost		$	$	$	$
Overhead	_% of prime cost				
Total Cost		$	$	$	$
Profit	_% of total cost				
Selling Price		$	$	$	$

Door Switch 15-Amp	#14-2	GROUPS			
		1	2	3	4
Total Prime Cost		$	$	$	$
Overhead	_% of prime cost				
Total Cost		$	$	$	$
Profit	_% of total cost				
Selling Price		$	$	$	$

Door Switch 20-Amp	#12-2	GROUPS			
		1	2	3	4
Total Prime Cost		$	$	$	$
Overhead	_% of prime cost				
Total Cost		$	$	$	$
Profit	_% of total cost				
Selling Price		$	$	$	$

Weatherproof Switch, 15-Amp	#14-2	GROUPS			
		1	2	3	4
Total Prime Cost		$	$	$	$
Overhead	_% of prime cost				
Total Cost		$	$	$	$
Profit	_% of total cost				
Selling Price		$	$	$	$

Weatherproof Switch, 20-Amp	#12-2	GROUPS			
		1	2	3	4
Total Prime Cost		$	$	$	$
Overhead	_% of prime cost				
Total Cost		$	$	$	$
Profit	_% of total cost				
Selling Price		$	$	$	$

Branch Circuit to Fixed Equipment, 15-Amp, 2-Wire	#14-2	GROUPS			
		1	2	3	4
Total Prime Cost		$	$	$	$
Overhead	_% of prime cost				
Total Cost		$	$	$	$
Profit	_% of total cost				
Selling Price		$	$	$	$

Branch Circuit to Fixed Equipment, 15-Amp, 3-Wire	#14-3	GROUPS			
		1	2	3	4
Total Prime Cost		$	$	$	$
Overhead	_% of prime cost				
Total Cost		$	$	$	$
Profit	_% of total cost				
Selling Price		$	$	$	$

Branch Circuit to Fixed Equipment, 20-Amp, 2-Wire	#12-2	GROUPS			
		1	2	3	4
Total Prime Cost		$	$	$	$
Overhead	_% of prime cost				
Total Cost		$	$	$	$
Profit	_% of total cost				
Selling Price		$	$	$	$

Branch Circuit to Fixed Equipment, 20-Amp, 3-Wire	#12-3	GROUPS			
		1	2	3	4
Total Prime Cost		$	$	$	$
Overhead	_% of prime cost				
Total Cost		$	$	$	$
Profit	_% of total cost				
Selling Price		$	$	$	$

Branch Circuit to Fixed Equipment, 30-Amp, 2-Wire	#10-2	GROUPS			
		1	2	3	4
Total Prime Cost		$	$	$	$
Overhead	_% of prime cost				
Total Cost		$	$	$	$
Profit	_% of total cost				
Selling Price		$	$	$	$

Branch Circuit to Fixed Equipment, 30-Amp, 3-Wire	#10-3	GROUPS			
		1	2	3	4
Total Prime Cost		$	$	$	$
Overhead	_% of prime cost				
Total Cost		$	$	$	$
Profit	_% of total cost				
Selling Price		$	$	$	$

Branch Circuit to Special Receptacle, 20-Amp, 3-Wire	#12-3	GROUPS			
		1	2	3	4
Total Prime Cost		$	$	$	$
Overhead	_% of prime cost				
Total Cost		$	$	$	$
Profit	_% of total cost				
Selling Price		$	$	$	$

Branch Circuit to Special Receptacle, 30-Amp, 3-Wire	#10-3	GROUPS			
		1	2	3	4
Total Prime Cost		$	$	$	$
Overhead	_% of prime cost				
Total Cost		$	$	$	$
Profit	_% of total cost				
Selling Price		$	$	$	$

Branch Circuit to Special Receptacle, 40-Amp, 3-Wire	#8-3	GROUPS			
		1	2	3	4
Total Prime Cost		$	$	$	$
Overhead	_% of prime cost				
Total Cost		$	$	$	$
Profit	_% of total cost				
Selling Price		$	$	$	$

Branch Circuit to Range Receptacle, 40-Amp, 3-Wire	#8-3	GROUPS			
		1	2	3	4
Total Prime Cost		$	$	$	$
Overhead	_% of prime cost				
Total Cost		$	$	$	$
Profit	_% of total cost				
Selling Price		$	$	$	$

Branch Circuit to Range Receptacle, 50-Amp, 3-Wire	#6-3	GROUPS			
		1	2	3	4
Total Prime Cost		$	$	$	$
Overhead	_% of prime cost				
Total Cost		$	$	$	$
Profit	_% of total cost				
Selling Price		$	$	$	$

Feeder, 240V, 3-Wire, 40-Amp	#8-3	GROUPS			
		1	2	3	4
Total Prime Cost		$	$	$	$
Overhead	_% of prime cost				
Total Cost		$	$	$	$
Profit	_% of total cost				
Selling Price		$	$	$	$

Feeder, 240V, 3-Wire, 60-Amp	#6-3	GROUPS			
		1	2	3	4
Total Prime Cost		$	$	$	$
Overhead	_% of prime cost				
Total Cost		$	$	$	$
Profit	_% of total cost				
Selling Price		$	$	$	$

Feeder, 240V, 3-Wire, 100-Amp	#3-3	GROUPS			
		1	2	3	4
Total Prime Cost		$	$	$	$
Overhead	_% of prime cost				
Total Cost		$	$	$	$
Profit	_% of total cost				
Selling Price		$	$	$	$

Feeder, 240V, 3-Wire, 125-Amp	#1-3	GROUPS			
		1	2	3	4
Total Prime Cost		$	$	$	$
Overhead	_% of prime cost				
Total Cost		$	$	$	$
Profit	_% of total cost				
Selling Price		$	$	$	$

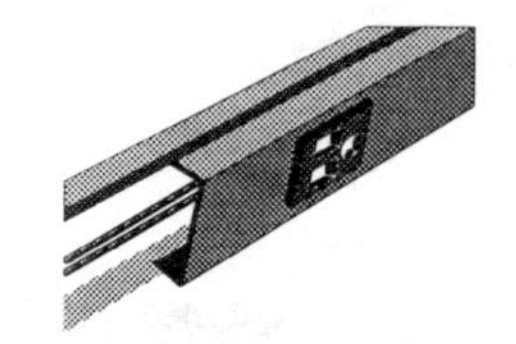

Chapter 4
Surface Metal Raceway

When it is impractical to install the wiring in concealed areas, surface metal raceway (Figures 4-1 through 4-3, beginning on the next page) is a good compromise. Even though it is visible, the proper painting of it to match the color of ceilings and walls makes it very inconspicuous. Surface metal raceway, traditionally, is made from sheet metal strips drawn into shape and comes in various shapes and sizes with factory fittings to meet nearly every application found in finished areas of residential buildings. A complete list of fittings can be obtained at your local electrical equipment supplier. In recent years, a nonmetallic version has become available, but the installation time and material costs are nearly the same.

The running of straight lines of surface molding is simple. A coupling is slipped in the end of a length of molding so that the coupling screw hole is exposed. The coupling may then be screwed to the surface to which the molding is to be attached. Then another length of molding is slipped on the opposite end of the coupling. This process is repeated until the installation is complete.

Factory fittings are used for corners and turns or the molding may be bent (to a certain extent) with a special bender. Matching outlet boxes for surface mounting are also available, and bushings are necessary at such boxes to prevent the sharp edges of the molding from injuring the insulation on the wire.

INSTALL 2000B BASE ON SURFACE.

Starting with feed section, mount entire run off 2000B base with No. 8 flat head screw through screw piercings and knockouts. Cut base to length at corners and end of run, as required.

BRING FEED INTO 2000B BASE.

Back-feed connection shown. See back of sheet for alternate methods of feeding.

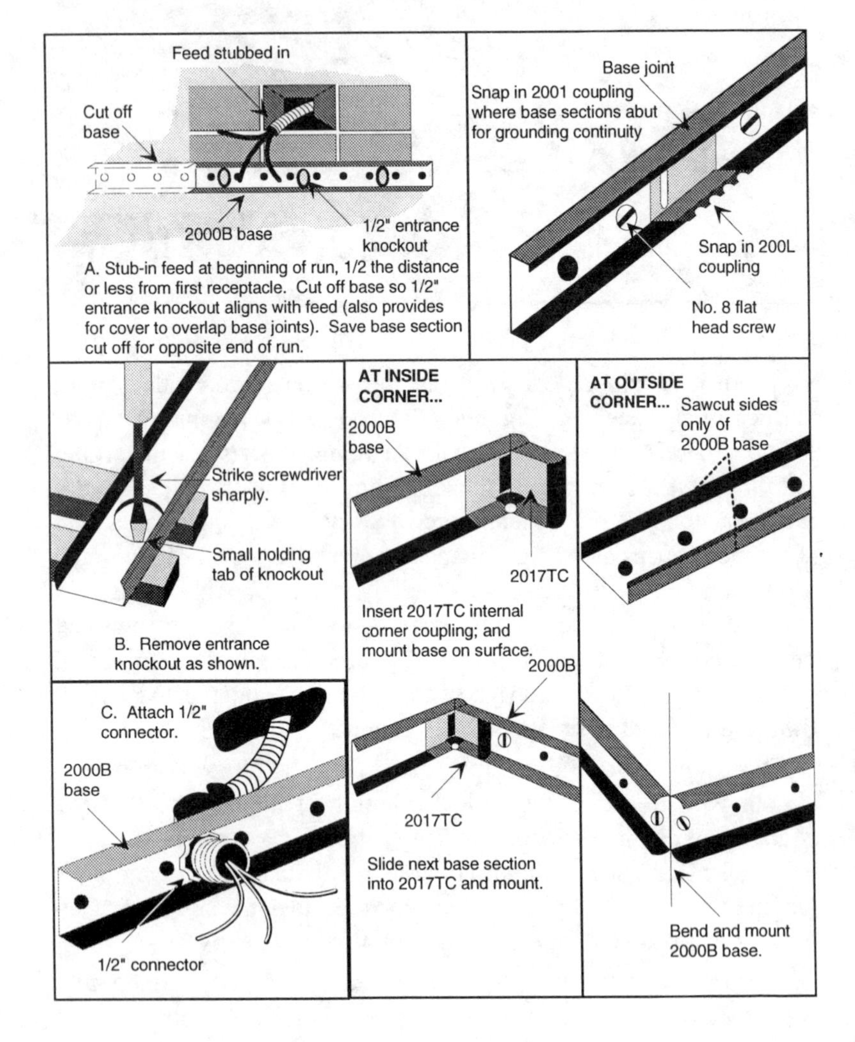

Figure 4-1: Basic installation methods for surface metal raceway.

CONNECT SNAPICOIL TO FEED.

Layout snapicoil along entire run of base so that receptacles are not located over feed or in corners. Connect to feed wires with W30 pressure type wire connectors (for common connection of 2, 3, or 4 No. 12 or No. 14 solid conductors); NOT TO BE USED to connect equipment grounding conductors. Insert only conductors of same color in a connector.

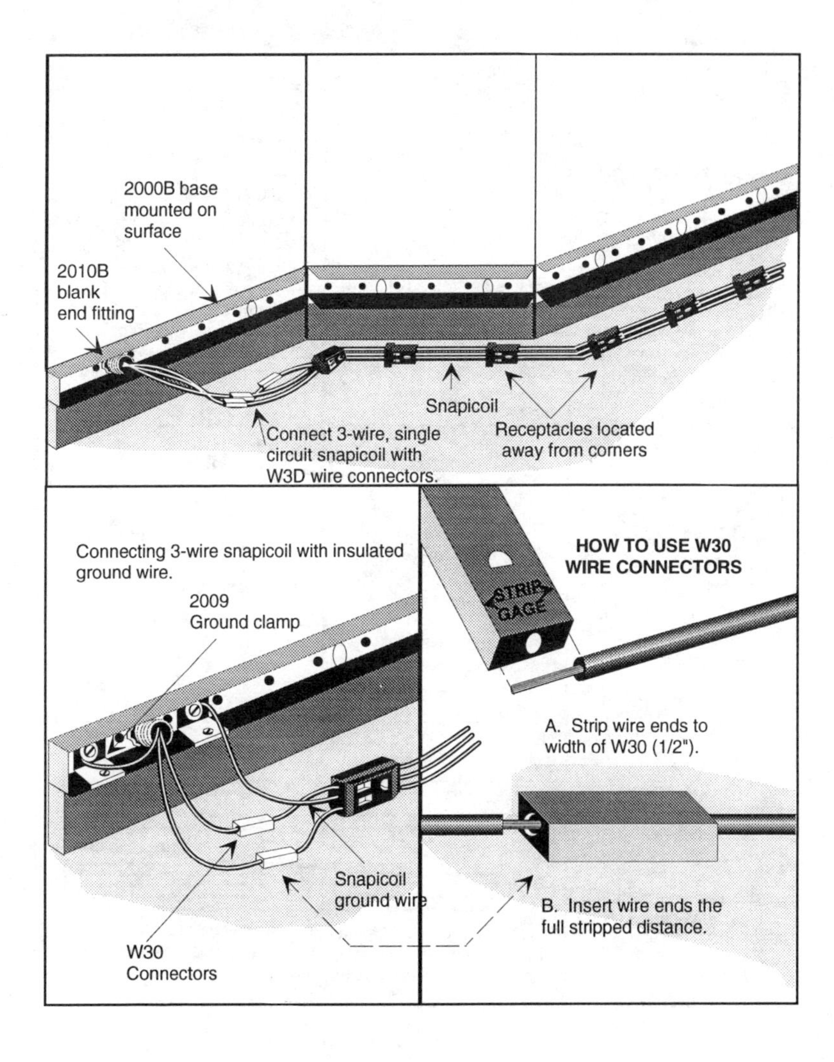

Figure 4-2: Connect snapicoil to feed.

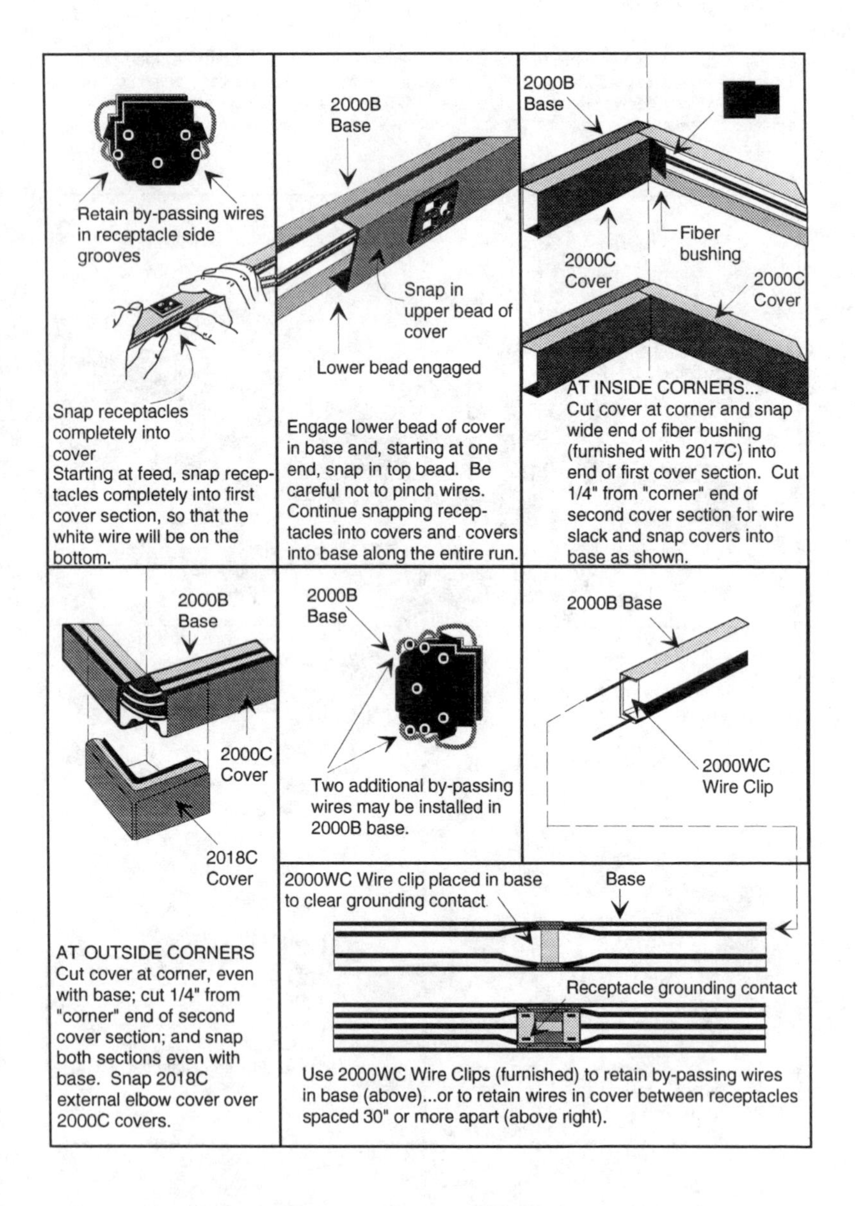

Figure 4-3: Assemble snapicoil and 2000C cover in 2000B base. Cover sections should overlap base joints for rigidity and better ground continuity.

Clips are used to fasten the molding in place. The clip is secured by a screw and then the molding is slipped into the clip. When parallel runs of molding are installed, they may be secured in place by means of a multiple strap. The joints in runs of molding are covered by slipping a connection cover over the joints. Such runs of molding should be grounded the same as any other metal raceway, and this is done by the use of grounding clips. The current-carrying wires are normally pulled in after the molding is in place.

The installation of surface metal molding requires no special tools unless bending the molding is necessary. The molding is fastened in place with screws, toggle bolts, and the like, depending on the materials to which it is fastened. All molding should be run straight and parallel with the room or building lines, that is, baseboards, trims, and other room moldings. The decor of the room should be considered first and the molding made as inconspicuous as possible.

It is often desirable to install surface molding not used for wires in order to complete a pattern set by other surface molding containing current-carrying wires, or to continue a run to make it appear to be part of the room's decoration.

The use of surface metal raceway is permitted in dry locations. It must not be used under the following conditions:

- Where subject to severe physical damage.
- Where the voltage is 300 volts or more between conductors unless the metal has a thickness of not less than .040 inch.
- Where subjected to corrosive vapors.
- In hoistways.
- In any hazardous location except Class I, Division 2 locations as permitted in the exception to *NEC* Section 501-4(b).
- In concealed areas, with some exceptions. Since this wiring system is not concealed, only three installation groups are included in the estimating charts.

Surface-Mounted Lighting Outlet

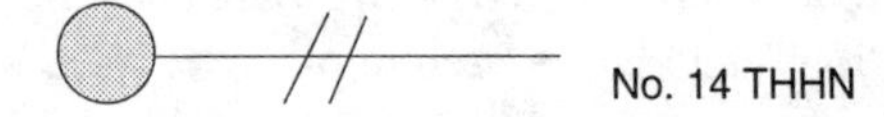

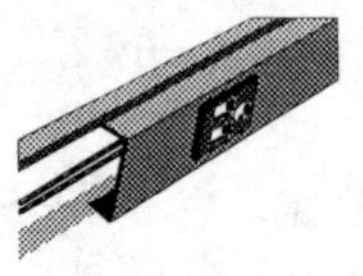

Cost Item	Quantity	Price or Rate	Per	Installation Groups 1	2	3
MATERIAL						
Wiremold fixture box	1		E			
#14 Type THHN building wire	34 ft.		C Ft.			
Surface metal raceway	15 ft.					
Wiremold fitting	1					
Miscellaneous	Lot					
TOTAL MATERIAL COST						
TOTAL LABOR COST - GROUP 1	1.55 WH		Hr			
GROUP 2	2.30 WH		Hr			
GROUP 3	3.10 WH		Hr			
DIRECT JOB EXPENSE						
TOTAL PRIME COST				$	$	$

Wall-Mounted Lighting Outlet

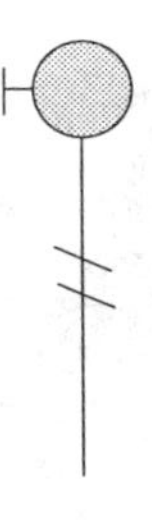

No. 14 THHN

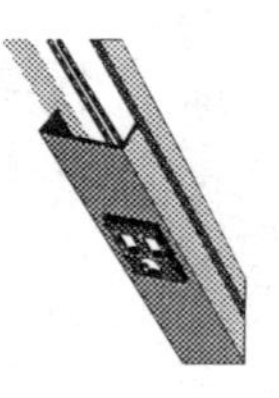

Cost Item	Quantity	Price or Rate	Per	Installation Groups		
MATERIAL				1	2	3
Wiremold fixture box	1		E			
#14 Type THHN building wire	34 ft.		C Ft.			
Surface metal raceway	15 ft.					
Wiremold fitting	1					
Miscellaneous	Lot					
TOTAL MATERIAL COST						
TOTAL LABOR COST - GROUP 1	1.45 WH		Hr			
GROUP 2	2.20 WH		Hr			
GROUP 3	3.00 WH		Hr			
DIRECT JOB EXPENSE						
TOTAL PRIME COST				$	$	$

Duplex Receptacle

No. 14 THHN

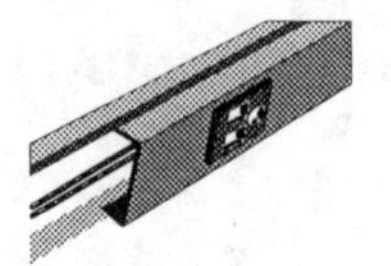

Cost Item	Quantity	Price or Rate	Per	Installation Groups 1	2	3
MATERIAL						
Wiremold receptacle box	1		E			
#14 Type THHN building wire	34 ft.		C Ft.			
Surface metal raceway	15 ft.		C Ft.			
Wiremold fitting	1		E			
Duplex receptacle and plate	1		E			
Miscellaneous	Lot					
TOTAL MATERIAL COST						
TOTAL LABOR COST - GROUP 1	1.65 WH		Hr			
GROUP 2	2.60 WH		Hr			
GROUP 3	3.50 WH		Hr			
DIRECT JOB EXPENSE						
TOTAL PRIME COST				$	$	$

Duplex Receptacle

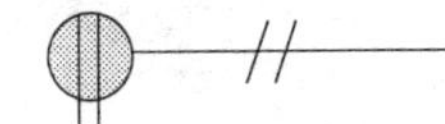

No. 12 THHN

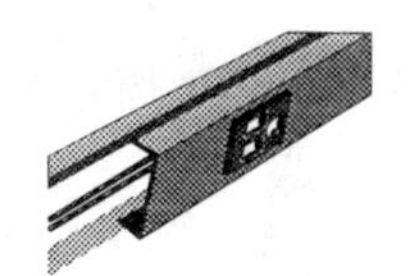

Cost Item	Quantity	Price or Rate	Per	Installation Groups 1	2	3
MATERIAL						
Wiremold receptacle box	1		E			
#12 Type THHN building wire	34 ft.		C Ft.			
Surface metal raceway	15 ft.		C Ft.			
Wiremold fitting	1		E			
Duplex receptacle and plate	1		E			
Miscellaneous	Lot					
TOTAL MATERIAL COST						
TOTAL LABOR COST - GROUP 1	1.75 WH		Hr			
GROUP 2	2.65 WH		Hr			
GROUP 3	3.60 WH		Hr			
DIRECT JOB EXPENSE						
TOTAL PRIME COST				$	$	$

Duplex Receptacle - Split-Wired

No. 14 THHN

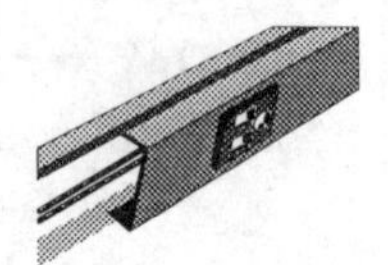

Cost Item	Quantity	Price or Rate	Per	Installation Groups 1	2	3
MATERIAL						
Wiremold receptacle box	1		E			
#14 Type THHN building wire	50 ft.		C Ft.			
Surface metal raceway	15 ft.		C Ft.			
Wiremold fitting	1		E			
Duplex receptacle and plate	1		E			
Miscellaneous	Lot					
TOTAL MATERIAL COST						
TOTAL LABOR COST - GROUP 1	1.80 WH		Hr			
GROUP 2	2.75 WH		Hr			
GROUP 3	3.65 WH		Hr			
DIRECT JOB EXPENSE						
TOTAL PRIME COST				$	$	$

Clock Receptacle

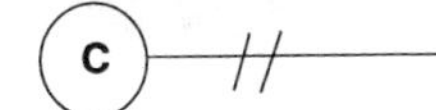

No. 14 THHN

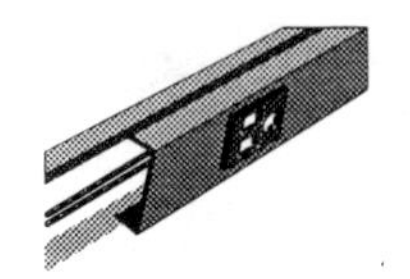

Cost Item	Quantity	Price or Rate	Per	Installation Groups		
				1	2	3
MATERIAL						
Wiremold receptacle box	1		E			
#14 Type THHN building wire	34 ft.		C Ft.			
Surface metal raceway	15 ft.		C Ft.			
Wiremold fittings	2		E			
Clock hanger receptacle and cover	1		E			
Miscellaneous	Lot					
TOTAL MATERIAL COST						
TOTAL LABOR COST - GROUP 1	1.80 WH		Hr			
GROUP 2	2.75 WH		Hr			
GROUP 3	3.65 WH		Hr			
DIRECT JOB EXPENSE						
TOTAL PRIME COST				$	$	$

Fan Receptacle

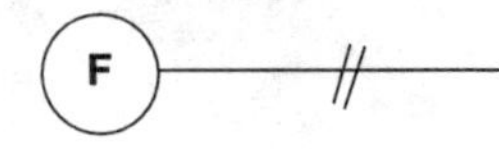

No. 12 THHN

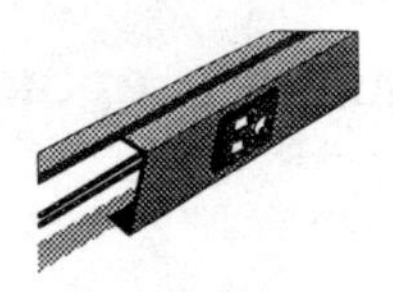

Cost Item	Quantity	Price or Rate	Per	Installation Groups 1	2	3
MATERIAL						
Wiremold ceiling-fan box	1		E			
#12 Type THHN building wire	35 ft.		C Ft.			
Surface metal raceway	15 ft.		C Ft.			
Wiremold fittings	2		E			
Miscellaneous	Lot					
TOTAL MATERIAL COST						
TOTAL LABOR COST - GROUP 1	1.85 WH		Hr			
GROUP 2	2.80 WH		Hr			
GROUP 3	3.65 WH		Hr			
DIRECT JOB EXPENSE						
TOTAL PRIME COST				$	$	$

Multioutlet Assembly - 3-Foot Section

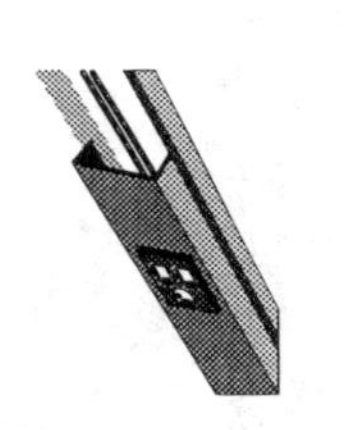

Cost Item	Quantity	Price or Rate	Per	Installation Groups		
MATERIAL				1	2	3
Plugmold, 3-foot section - prewired	1		E			
Plugmold fitting or adapter	1		E			
Miscellaneous	Lot					
TOTAL MATERIAL COST						
TOTAL LABOR COST - GROUP 1	1.70 WH		Hr			
GROUP 2	2.65 WH		Hr			
GROUP 3	3.50 WH		Hr			
DIRECT JOB EXPENSE						
TOTAL PRIME COST				$	$	$

Multioutlet Assembly - 4-Foot Section

No. 12 THHN

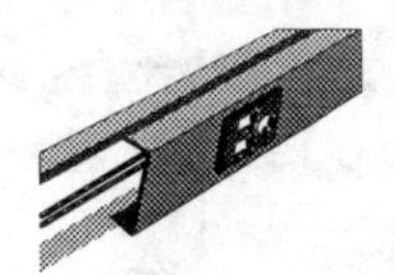

Cost Item	Quantity	Price or Rate	Per	Installation Groups 1	2	3
MATERIAL						
Plugmold, 4-foot section - prewired	1		E			
Plugmold fitting or adapter	1		E			
Miscellaneous	Lot					
TOTAL MATERIAL COST						
TOTAL LABOR COST - GROUP 1	1.80 WH		Hr			
GROUP 2	2.85 WH		Hr			
GROUP 3	3.80 WH		Hr			
DIRECT JOB EXPENSE						
TOTAL PRIME COST				$	$	$

Multioutlet Assembly - 5-Foot Section

No. 12 THHN

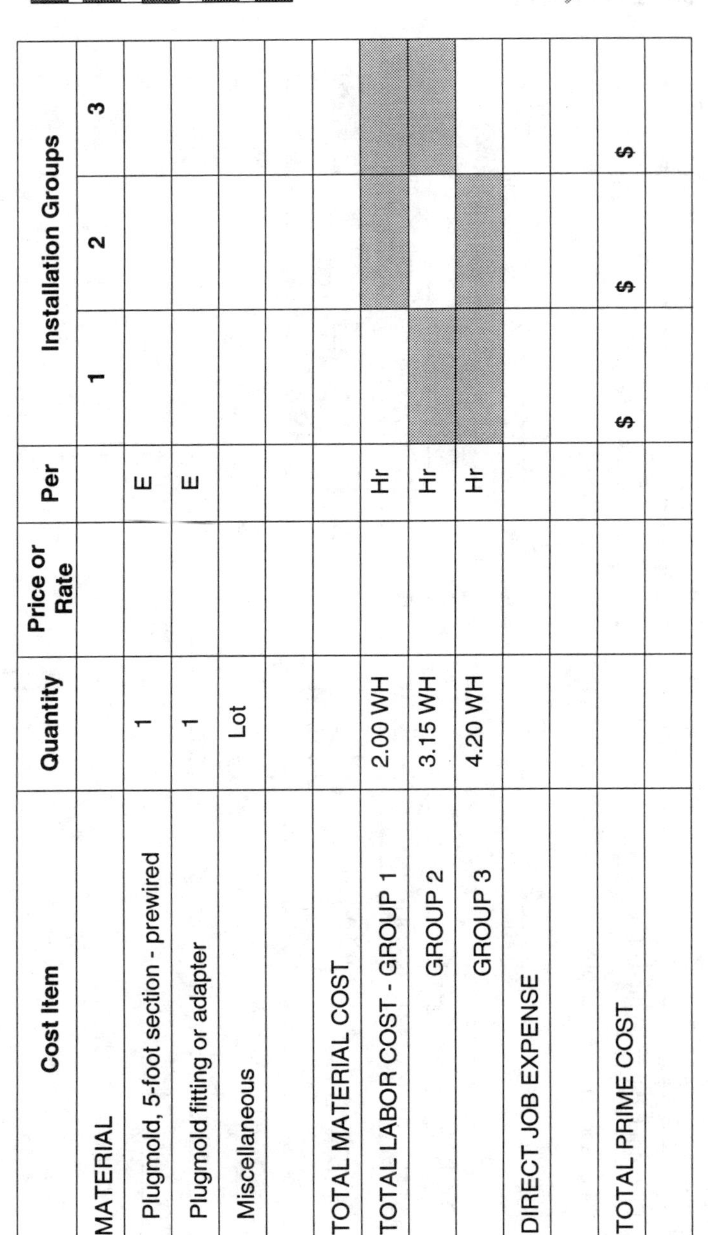

Cost Item	Quantity	Price or Rate	Per	Installation Groups 1	2	3
MATERIAL						
Plugmold, 5-foot section - prewired	1		E			
Plugmold fitting or adapter	1		E			
Miscellaneous	Lot					
TOTAL MATERIAL COST						
TOTAL LABOR COST - GROUP 1	2.00 WH		Hr			
GROUP 2	3.15 WH		Hr			
GROUP 3	4.20 WH		Hr			
DIRECT JOB EXPENSE						
TOTAL PRIME COST				$	$	$

Multioutlet Assembly - 6-Foot Section

No. 12 THHN

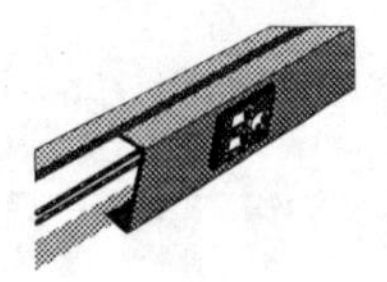

Cost Item	Quantity	Price or Rate	Per	Installation Groups 1	2	3
MATERIAL						
Plugmold, 6-foot section - prewired	1		E			
Plugmold fitting or adapter	1		E			
Miscellaneous	Lot					
TOTAL MATERIAL COST						
TOTAL LABOR COST - GROUP 1	2.30 WH		Hr			
GROUP 2	3.25 WH		Hr			
GROUP 3	4.50 WH		Hr			
DIRECT JOB EXPENSE						
TOTAL PRIME COST				$	$	$

Wall Switch - Single-Pole

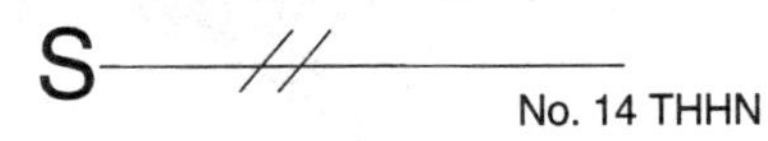

Cost Item	Quantity	Price or Rate	Per	Installation Groups 1	2	3
MATERIAL						
Wiremold switch box	1		E			
Wiremold	15 ft.		C ft.			
Wiremold fittings	2		E			
#14 THHN wire	32 ft.					
Single-pole switch and plate	1		E			
Miscellaneous	Lot					
TOTAL MATERIAL COST						
TOTAL LABOR COST - GROUP 1	1.65 WH		Hr			
GROUP 2	2.50 WH		Hr			
GROUP 3	3.30 WH		Hr			
DIRECT JOB EXPENSE						
TOTAL PRIME COST				$	$	$

Wall Switch with Pilot Light

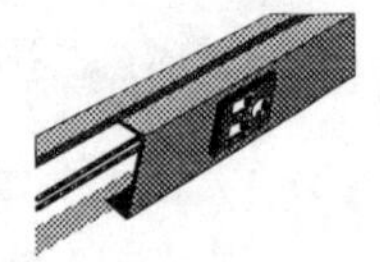

Cost Item	Quantity	Price or Rate	Per	Installation Groups 1	2	3
MATERIAL						
Wiremold switch box	1		E			
Wiremold	15 ft.		C ft.			
Wiremold fittings	2		E			
#14 THHN wire	47 ft.		M ft.			
Switch with pilot light and cover	1		E			
Miscellaneous	Lot					
TOTAL MATERIAL COST						
TOTAL LABOR COST - GROUP 1	1.75 WH		Hr			
GROUP 2	2.60 WH		Hr			
GROUP 3	3.50 WH		Hr			
DIRECT JOB EXPENSE						
TOTAL PRIME COST				$	$	$

Wall Switch - Three-Way

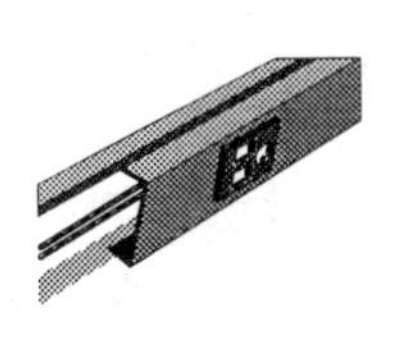

Cost Item	Quantity	Price or Rate	Per	Installation Groups 1	2	3
MATERIAL						
Wiremold switch box	1		E			
Wiremold	20 ft.		C ft.			
Wiremold fittings	2		E			
#14 THHN wire	65 ft.		M ft.			
Three-way wall switch and plate	1		E			
Miscellaneous	Lot					
TOTAL MATERIAL COST						
TOTAL LABOR COST - GROUP 1	2.30 WH		Hr			
GROUP 2	3.25 WH		Hr			
GROUP 3	4.50 WH		Hr			
DIRECT JOB EXPENSE						
TOTAL PRIME COST				$	$	$

Wall Switch - Four-Way

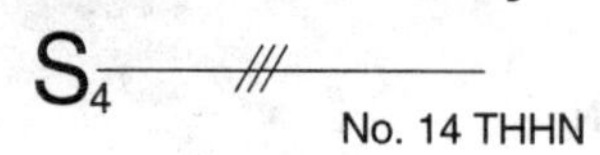

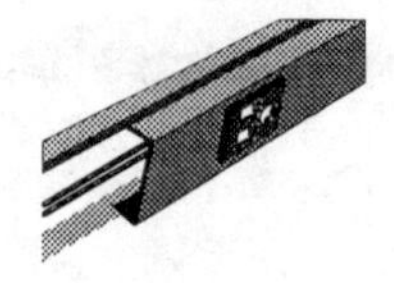

Cost Item	Quantity	Price or Rate	Per	Installation Groups		
MATERIAL				1	2	3
Wiremold switch box	1		E			
Wiremold	30 ft.		C ft.			
Wiremold fittings	2		E			
#14 THHN wire	92 ft.		M ft.			
Four-way switch and plate	1		E			
Miscellaneous	Lot					
TOTAL MATERIAL COST						
TOTAL LABOR COST - GROUP 1	2.15 WH		Hr			
GROUP 2	3.25 WH		Hr			
GROUP 3	4.35 WH		Hr			
DIRECT JOB EXPENSE						
TOTAL PRIME COST				$	$	$

Branch Circuit to Junction Box

No. 14 THHN

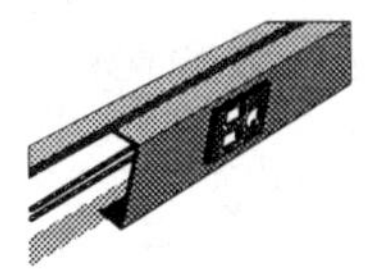

Cost Item	Quantity	Price or Rate	Per	Installation Groups 1	2	3
MATERIAL						
Wiremold junction box with cover	1		E			
Wiremold	15 ft.		C ft.			
Wiremold fittings	2		E			
#14 THHN wire	32 ft.		M ft.			
Miscellaneous	Lot					
TOTAL MATERIAL COST						
TOTAL LABOR COST - GROUP 1	1.90 WH		Hr			
GROUP 2	2.85 WH		Hr			
GROUP 3	3.90 WH		Hr			
DIRECT JOB EXPENSE						
TOTAL PRIME COST				$	$	$

Branch Circuit to Juntion Box

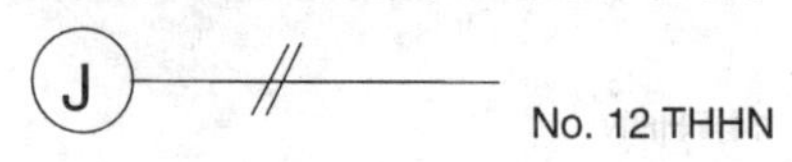

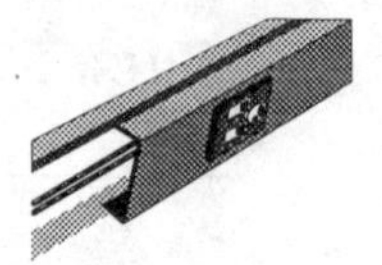

Cost Item	Quantity	Price or Rate	Per	Installation Groups 1	2	3
MATERIAL						
Wiremold junction box with cover	1		E			
Wiremold	15 ft.		C ft.			
Wiremold fittings	2		E			
#12 THHN wire	32 ft.		M ft.			
Miscellaneous	Lot					
TOTAL MATERIAL COST						
TOTAL LABOR COST - GROUP 1	2.05 WH		Hr			
GROUP 2	3.10 WH		Hr			
GROUP 3	4.15 WH		Hr			
DIRECT JOB EXPENSE						
TOTAL PRIME COST				$	$	$

Branch Circuit to Junction Box

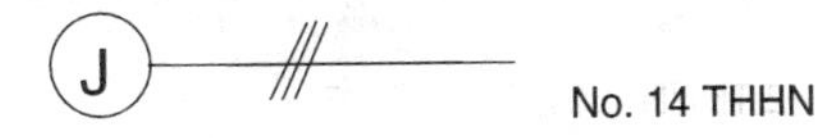

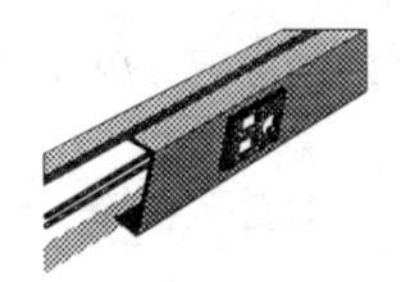

Cost Item	Quantity	Price or Rate	Per	Installation Groups 1	2	3
MATERIAL						
Wiremold junction box with cover	1		E			
Wiremold	15 ft.		C ft.			
Wiremold fittings	2		E			
#14 THHN wire	47 ft.		M ft.			
Miscellaneous	Lot					
TOTAL MATERIAL COST						
TOTAL LABOR COST - GROUP 1	2.10 WH		Hr			
GROUP 2	3.10 WH		Hr			
GROUP 3	4.10 WH		Hr			
DIRECT JOB EXPENSE						
TOTAL PRIME COST				$	$	$

Branch Circuit to Junction Box

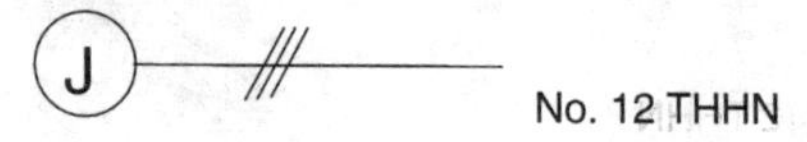

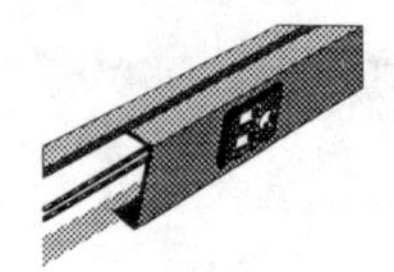

Cost Item	Quantity	Price or Rate	Per	Installation Groups 1	2	3
MATERIAL						
Wiremold junction box with cover	1		E			
Wiremold	15 ft.		C ft.			
Wiremold fittings	2		E			
#12 THHN wire	47 ft.		M ft.			
Miscellaneous	Lot					
TOTAL MATERIAL COST						
TOTAL LABOR COST - GROUP 1	2.20 WH		Hr			
GROUP 2	3.25 WH		Hr			
GROUP 3	4.30 WH		Hr			
DIRECT JOB EXPENSE						
TOTAL PRIME COST				$	$	$

Special Receptacle - 20A, 2-Wire

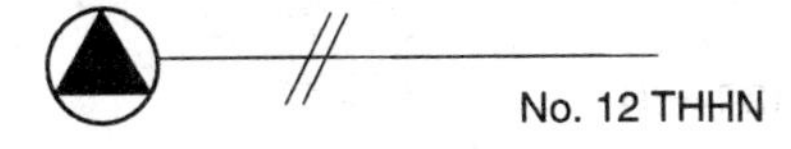

Cost Item	Quantity	Price or Rate	Per	Installation Groups 1	2	3
MATERIAL						
Wiremold receptacle	1		E			
Wiremold	15 ft.		C ft.			
Wiremold fittings	3		E			
#12 THHN wire	32 ft.		M ft.			
20-amp special-purpose receptacle	1		E			
Miscellaneous	Lot					
TOTAL MATERIAL COST						
TOTAL LABOR COST - GROUP 1	2.35 WH		Hr			
GROUP 2	3.55 WH		Hr			
GROUP 3	4.75 WH		Hr			
DIRECT JOB EXPENSE						
TOTAL PRIME COST				$	$	$

Special Receptacle - 30A, 3-Wire

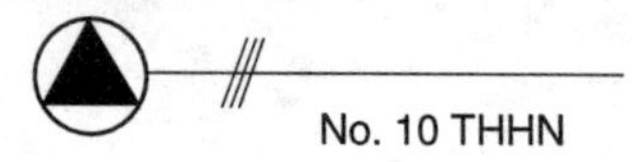

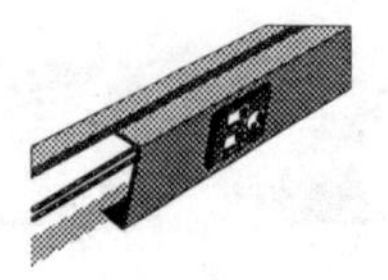

Cost Item	Quantity	Price or Rate	Per	Installation Groups 1	2	3
MATERIAL						
Wiremold receptacle	1		E			
Wiremold	15 ft.		C ft.			
Wiremold fittings	3		E			
#10 THHN wire	47 ft.		M ft.			
30-amp special-purpose receptacle	1		E			
Miscellaneous	Lot					
TOTAL MATERIAL COST						
TOTAL LABOR COST - GROUP 1	2.65 WH		Hr			
GROUP 2	4.55 WH		Hr			
GROUP 3	6.75 WH		Hr			
DIRECT JOB EXPENSE						
TOTAL PRIME COST				$	$	$

Range Receptacle

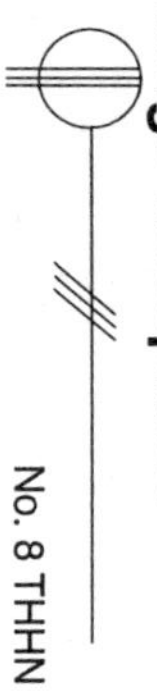

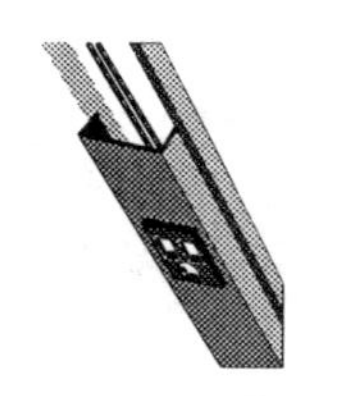

Cost Item	Quantity	Price or Rate	Per	Installation Groups		
MATERIAL				1	2	3
Wiremold receptacle box	1		E			
Wiremold	15 ft.		C ft.			
Wiremold fittings	3		E			
#8 THHN wire	50 ft.		M ft.			
Range receptacle	1		E			
Miscellaneous	Lot					
TOTAL MATERIAL COST						
TOTAL LABOR COST - GROUP 1	4.35 WH		Hr			
GROUP 2	5.55 WH		Hr			
GROUP 3	6.75 WH		Hr			
DIRECT JOB EXPENSE						
TOTAL PRIME COST				$	$	$

Selling-Price Tables

The selling-price tables to follow are designed for quick unit pricing of the various outlets found in this chapter, using surface metal raceway. Before arriving at a selling price, the preceding tables must be completed (filled in with appropriate prices). Then the prime cost determined for each wiring situation in the preceding tables should be entered in the corresponding selling-price tables to follow. Once the prime cost has been entered in each selling-price table, overhead and profit factors are calculated to arrive at a total unit selling price as described in detail in Chapter 2.

Surface-Mounted Lighting Outlet	**#14 THHN**	**GROUPS**		
		1	**2**	**3**
Total Prime Cost		**$**	**$**	**$**
Overhead	_% of prime cost			
Total Cost		**$**	**$**	**$**
Profit	_% of total cost			
Selling Price		**$**	**$**	**$**

Wall-Mounted Lighting Outlet	**#14 THHN**	**GROUPS**		
		1	**2**	**3**
Total Prime Cost		**$**	**$**	**$**
Overhead	_% of prime cost			
Total Cost		**$**	**$**	**$**
Profit	_% of total cost			
Selling Price		**$**	**$**	**$**

Duplex Receptacle 15-amp	**#14 THHN**	**GROUPS**		
		1	**2**	**3**
Total Prime Cost		**$**	**$**	**$**
Overhead	_% of prime cost			
Total Cost		**$**	**$**	**$**
Profit	_% of total cost			
Selling Price		**$**	**$**	**$**

Duplex Receptacle 20-amp	**#12 THHN**	**GROUPS**		
		1	**2**	**3**
Total Prime Cost		**$**	**$**	**$**
Overhead	_% of prime cost			
Total Cost		**$**	**$**	**$**
Profit	_% of total cost			
Selling Price		**$**	**$**	**$**

Duplex Receptacle Split-Wired	#14 THHN	GROUPS		
		1	2	3
Total Prime Cost		$	$	$
Overhead	_% of prime cost			
Total Cost		$	$	$
Profit	_% of total cost			
Selling Price		$	$	$

Clock Receptacle	#14 THHN	GROUPS		
		1	2	3
Total Prime Cost		$	$	$
Overhead	_% of prime cost			
Total Cost		$	$	$
Profit	_% of total cost			
Selling Price		$	$	$

Fan Receptacle	#12 THHN	GROUPS		
		1	2	3
Total Prime Cost		$	$	$
Overhead	_% of prime cost			
Total Cost		$	$	$
Profit	_% of total cost			
Selling Price		$	$	$

Multioutlet Assembly, 3 Feet	#12 THHN	GROUPS		
		1	2	3
Total Prime Cost		$	$	$
Overhead	_% of prime cost			
Total Cost		$	$	$
Profit	_% of total cost			
Selling Price		$	$	$

Multioutlet Assembly, 4 Feet	**#12 THHN**	**GROUPS**		
		1	**2**	**3**
Total Prime Cost		**$**	**$**	**$**
Overhead	_% of prime cost			
Total Cost		**$**	**$**	**$**
Profit	_% of total cost			
Selling Price		**$**	**$**	**$**

Multioutlet Assembly, 5 Feet	**#12 THHN**	**GROUPS**		
		1	**2**	**3**
Total Prime Cost		**$**	**$**	**$**
Overhead	_% of prime cost			
Total Cost		**$**	**$**	**$**
Profit	_% of total cost			
Selling Price		**$**	**$**	**$**

Multioutlet Assembly, 6 Feet	**#12 THHN**	**GROUPS**		
		1	**2**	**3**
Total Prime Cost		**$**	**$**	**$**
Overhead	_% of prime cost			
Total Cost		**$**	**$**	**$**
Profit	_% of total cost			
Selling Price		**$**	**$**	**$**

Wall Switch Single-Pole	**#14 THHN**	**GROUPS**		
		1	**2**	**3**
Total Prime Cost		**$**	**$**	**$**
Overhead	_% of prime cost			
Total Cost		**$**	**$**	**$**
Profit	_% of total cost			
Selling Price		**$**	**$**	**$**

Wall Switch With Pilot Light	#14 THHN	GROUPS		
		1	2	3
Total Prime Cost		$	$	$
Overhead	_% of prime cost			
Total Cost		$	$	$
Profit	_% of total cost			
Selling Price		$	$	$

Wall Switch Three-Way	#14 THHN	GROUPS		
		1	2	3
Total Prime Cost		$	$	$
Overhead	_% of prime cost			
Total Cost		$	$	$
Profit	_% of total cost			
Selling Price		$	$	$

Wall Switch Four-Way	#14 THHN	GROUPS		
		1	2	3
Total Prime Cost		$	$	$
Overhead	_% of prime cost			
Total Cost		$	$	$
Profit	_% of total cost			
Selling Price		$	$	$

Branch Circuit to Junction Box, 15-amp	#14 THHN	GROUPS		
		1	2	3
Total Prime Cost		$	$	$
Overhead	_% of prime cost			
Total Cost		$	$	$
Profit	_% of total cost			
Selling Price		$	$	$

Circuit to Junction Box, 20A, 2-Wire	#12 THHN	GROUPS		
		1	2	3
Total Prime Cost		$	$	$
Overhead	_% of prime cost			
Total Cost		$	$	$
Profit	_% of total cost			
Selling Price		$	$	$

Circuit to Junction Box, 15A, 3-Wire	#14 THHN	GROUPS		
		1	2	3
Total Prime Cost		$	$	$
Overhead	_% of prime cost			
Total Cost		$	$	$
Profit	_% of total cost			
Selling Price		$	$	$

Circuit to Junction Box, 20A, 3-Wire	#12 THHN	GROUPS		
		1	2	3
Total Prime Cost		$	$	$
Overhead	_% of prime cost			
Total Cost		$	$	$
Profit	_% of total cost			
Selling Price		$	$	$

Special Receptacle 20A, 2-Wire	#12 THHN	GROUPS		
		1	2	3
Total Prime Cost		$	$	$
Overhead	_% of prime cost			
Total Cost		$	$	$
Profit	_% of total cost			
Selling Price		$	$	$

Special Receptacle 30A, 3-Wire	**#10 THHN**	**GROUPS**		
		1	**2**	**3**
Total Prime Cost		$	$	$
Overhead	_% of prime cost			
Total Cost		$	$	$
Profit	_% of total cost			
Selling Price		$	$	$

Range Receptacle 40A, 3-Wire	**#8 THHN**	**GROUPS**		
		1	**2**	**3**
Total Prime Cost		$	$	$
Overhead	_% of prime cost			
Total Cost		$	$	$
Profit	_% of total cost			
Selling Price		$	$	$

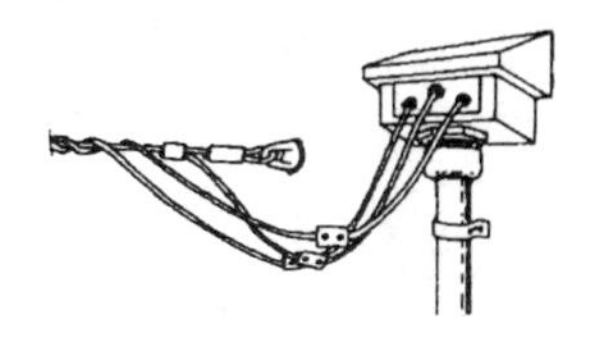

Chapter 5
Electric Services

An electric service enables the passage of electrical energy from the power company's lines to points of use within buildings. Figure 5-1 on the next page shows the basic sections of a residential electric service. Note that the high-voltage lines terminate on a power pole near the building that is being served. A transformer is mounted on the pole to reduce the voltage to a usable level of 120/240 volts, single-phase, 3-wire. The remaining sections are described as follows:

- *Service drop:* The overhead conductors, through which electrical service is supplied, between the last power company pole and the point of their connection to the service facilities located at the building or other support used for the purpose. In most cases, this part of the electric service is the responsibility of the power company.

- *Service entrance:* All components between the point of termination of the overhead service drop or underground service lateral and the building's main disconnecting device, except for the power company's metering equipment. All of these components are usually the responsibility of the electric contractor.

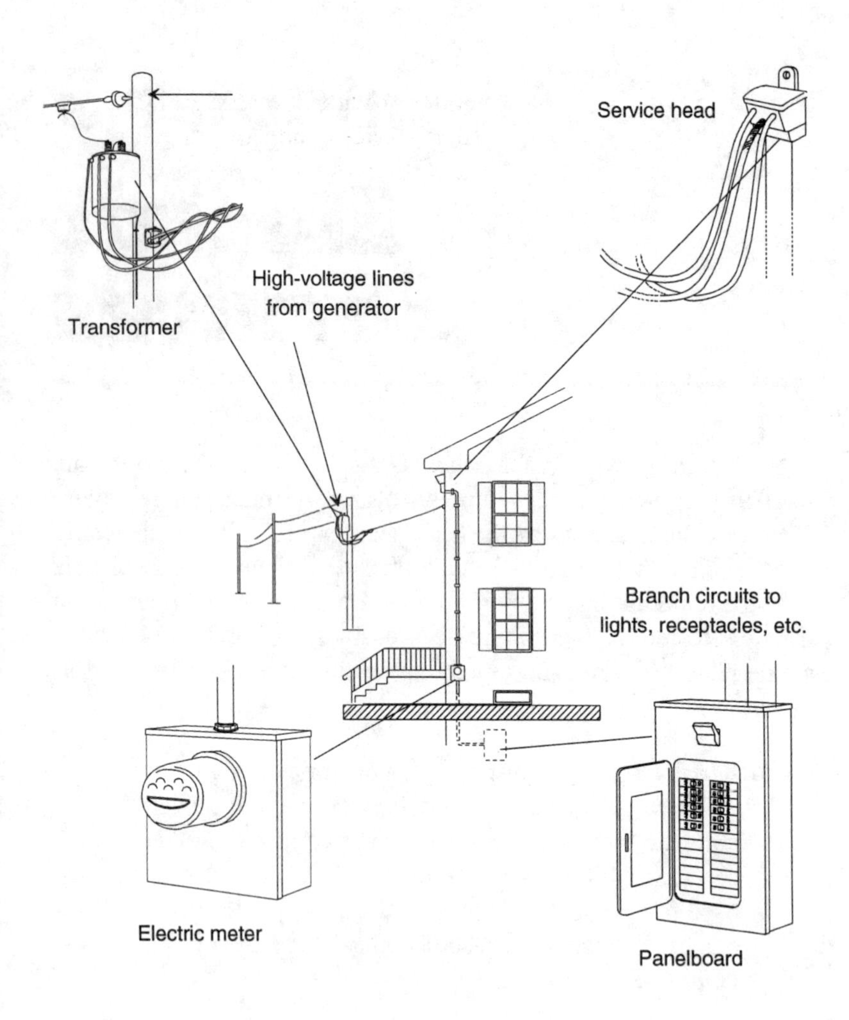

Figure 5-1: Components of a residential electric service.

- *Service-entrance conductors:* The conductors between the point of termination of the overhead service drop or underground service lateral and the main disconnecting device in the building.

- *Service-entrance equipment:* Provides overcurrent protection to the feeder and service conductors, a means of disconnecting the feeders from energized service conductors, and a means of measuring the energy used by means of metering equipment.

When the service conductors to the building are routed underground, as shown in Figure 5-2 on the next page, these conductors are known as the service lateral, defined as follows:

- *Service lateral:* The underground conductors through which electric service is supplied between the power company's distribution facilities and the first point of their connection to the building or area other support used for the purpose.

The estimating tables in this chapter are divided into three parts:

- Service-entrance conductors

- Service-entrance equipment

- Service grounding

The way in which these tables are arranged should provide all the necessary information required to estimate the selling price for the majority of all residential installations.

Types Of Services

The wiring method used for residential electrical services varies with each electrical contractor and with each area or juridiction in which the house is built. The least expensive method is service-entrance cable (Type S.E.) with aluminum conductors. However, this wiring method is not allowed in some areas, especially those subjected to heavy winter ice and snow conditions. The use of copper conductors upgrades the system slightly, but labor is essentially the same and no mechanical-protection advantage is derived.

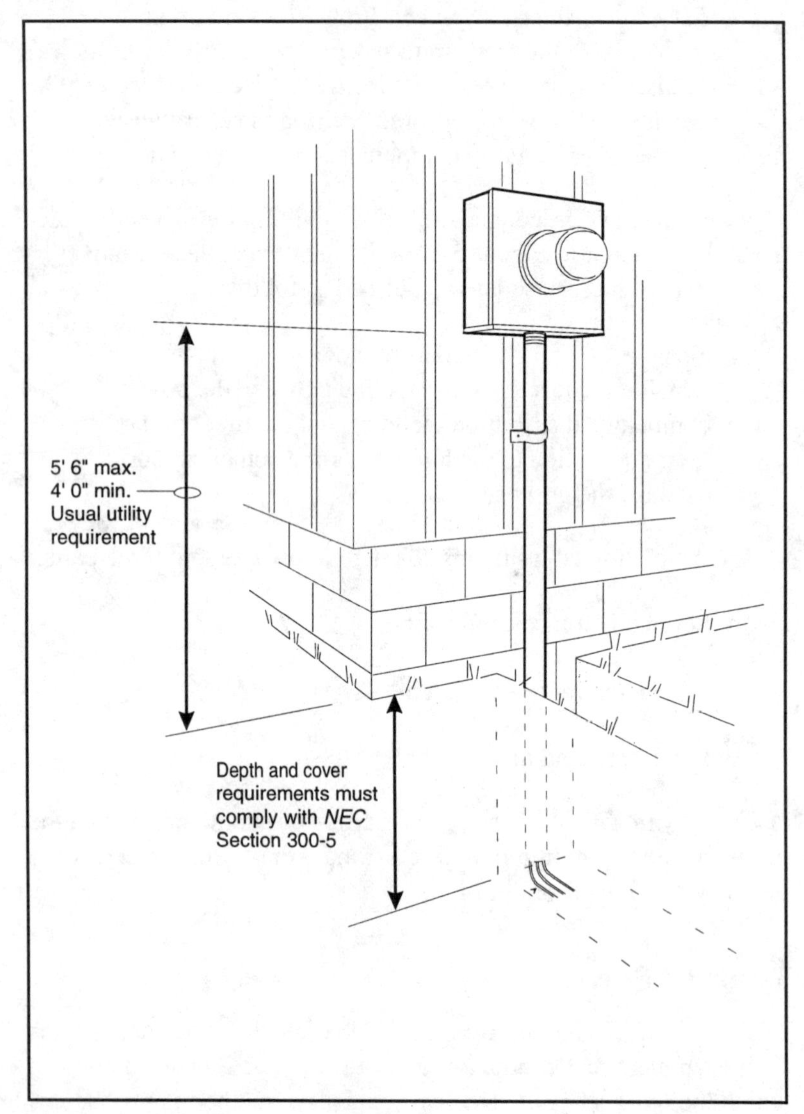

Figure 5-2: Underground service lateral.

The use of a conduit system — either rigid steel, rigid nonmetallic, EMT, aluminum, etc. — increases the stability of any service entrance, and is mandatory when using a mast-through-roof service as shown in Figure 5-3.

The components of each type of service-entrance are shown in the following illustrations beginning with Figure 5-4.

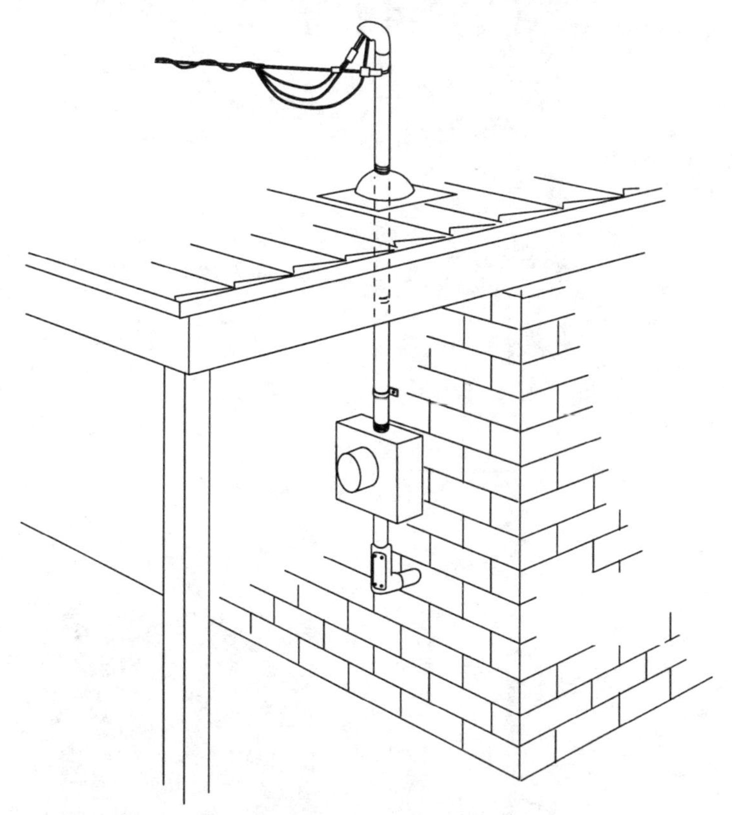

Figure 5-3: A rigid service mast must be used when the service-entrance raceway protrudes through a roof.

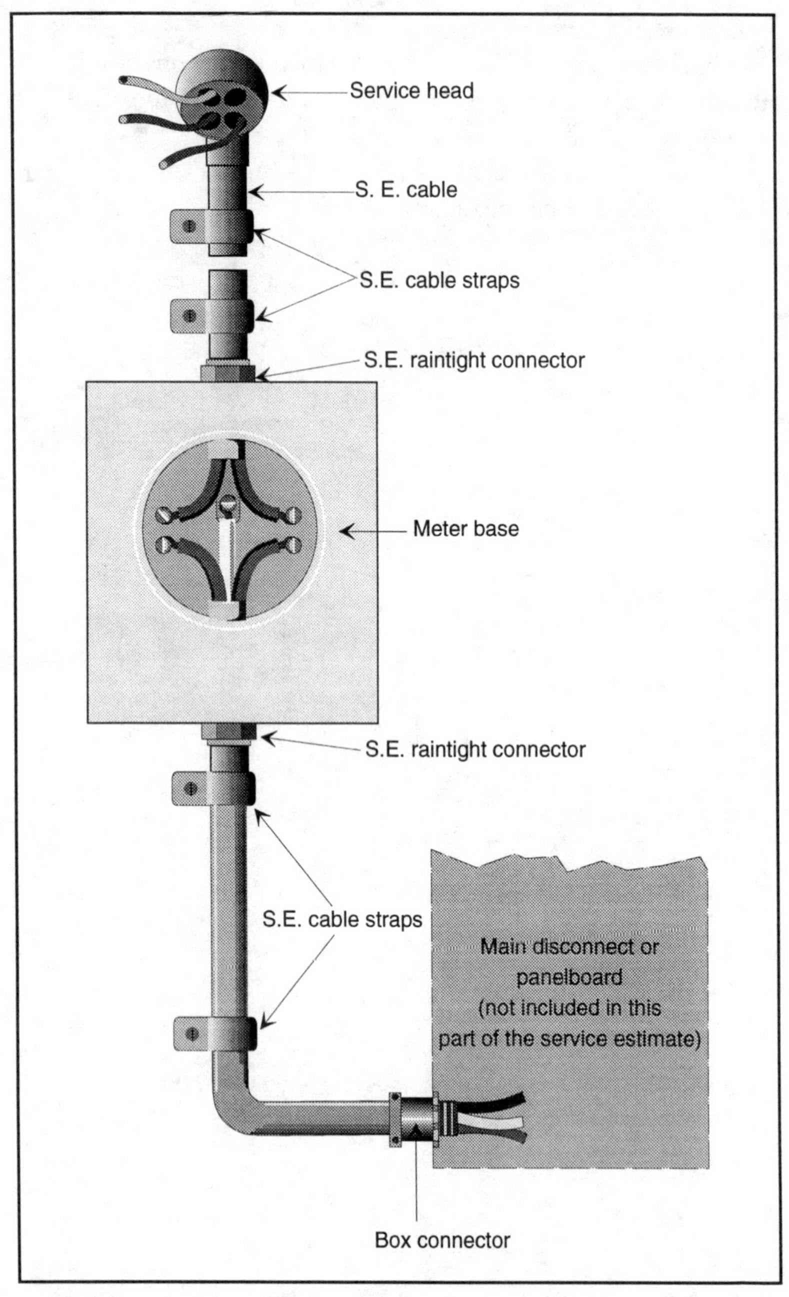

Figure 5-4: Components of an aluminum service-entrance cable installation.

Service-Entrance, 100A, Aluminum S.E. Cable

No. 2 Wire

Cost Item	Quantity	Price or Rate	Per	Installation Groups			
MATERIAL				1	2	3	4
Service head	1		E				
Raintight S.E. connectors	2		E				
S.E. connector	1		E				
Meter base	1		E				
#2/3 S.E. cable	25 ft.		C Ft.				
Miscellaneous	Lot						
TOTAL MATERIAL COST							
TOTAL LABOR COST - GROUP 1	5.55 WH		Hr				
GROUP 2	8.85 WH		Hr				
GROUP 3	11.15 WH		Hr				
GROUP 4	14.45 WH		Hr				
DIRECT JOB EXPENSE							
TOTAL PRIME COST				$	$	$	$

Service-Entrance, 125A, Aluminum S.E. Cable

No. 1/0 Wire

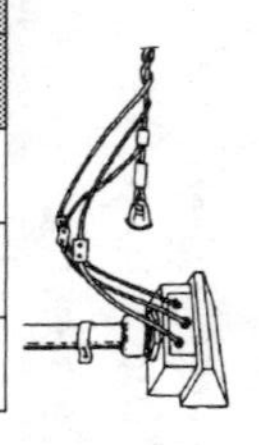

Cost Item	Quantity	Price or Rate	Per	Installation Groups			
				1	2	3	4
MATERIAL							
Service head	1		E				
Raintight S.E. connectors	2		E				
S.E. connector	1		E				
Meter base	1		E				
#1/0-3 S.E. cable	25 ft.		C Ft.				
Miscellaneous	Lot						
TOTAL MATERIAL COST							
TOTAL LABOR COST - GROUP 1	6.00 WH		Hr				
GROUP 2	9.00 WH		Hr				
GROUP 3	12.00 WH		Hr				
GROUP 4	15.00 WH		Hr				
DIRECT JOB EXPENSE							
TOTAL PRIME COST				$	$	$	$

Service-Entrance, 150A, Aluminum S.E. Cable

No. 2/0 Wire

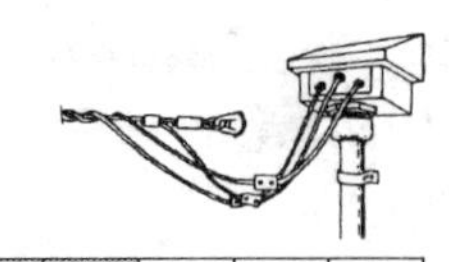

Cost Item	Quantity	Price or Rate	Per	Installation Groups 1	2	3	4
MATERIAL							
Service head	1		E				
Raintight S.E. connectors	2		E				
S.E. connector	1		E				
Meter base	1		E				
#2/0-3 S.E. cable	25 ft.		C Ft.				
Miscellaneous	Lot						
TOTAL MATERIAL COST							
TOTAL LABOR COST - GROUP 1	7.00 WH		Hr				
GROUP 2	10.00 WH		Hr				
GROUP 3	13.00 WH		Hr				
GROUP 4	16.00 WH		Hr				
DIRECT JOB EXPENSE							
TOTAL PRIME COST				$	$	$	$

Service-Entrance, 200A, Aluminum S.E. Cable

No. 4/0 Wire

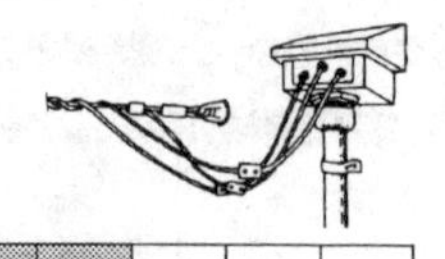

Cost Item	Quantity	Price or Rate	Per	Installation Groups 1	2	3	4
MATERIAL							
Service head	1		E				
Raintight S.E. connectors	2		E				
S.E. connector	1		E				
Meter base	1		E				
#4/0-3 S.E. cable	25 ft.		C Ft.				
Miscellaneous	Lot						
TOTAL MATERIAL COST							
TOTAL LABOR COST - GROUP 1	8.00 WH		Hr				
GROUP 2	11.00 WH		Hr				
GROUP 3	14.50 WH		Hr				
GROUP 4	17.00 WH		Hr				
DIRECT JOB EXPENSE							
TOTAL PRIME COST				$	$	$	$

Service-Entrance, 100A, Copper S.E. Cable

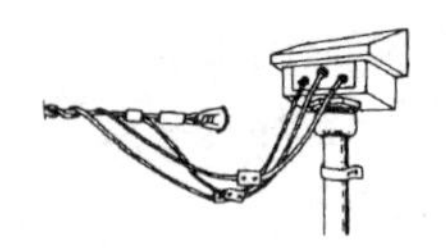

No. 4 Wire

Cost Item	Quantity	Price or Rate	Per	Installation Groups 1	2	3	4
MATERIAL							
Service head	1		E				
Raintight S.E. connectors	2		E				
S.E. connector	1		E				
Meter base	1		E				
#4-3 S.E. cable	25 ft.		C Ft.				
Miscellaneous	Lot						
TOTAL MATERIAL COST							
TOTAL LABOR COST - GROUP 1	6.00 WH		Hr				
GROUP 2	9.00 WH		Hr				
GROUP 3	12.50 WH		Hr				
GROUP 4	15.00 WH		Hr				
DIRECT JOB EXPENSE							
TOTAL PRIME COST				$	$	$	$

Service-Entrance, 125A, Copper S.E. Cable

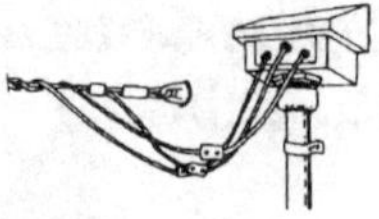

No. 2 Wire

Cost Item	Quantity	Price or Rate	Per	Installation Groups 1	2	3	4
MATERIAL							
Service head	1		E				
Raintight S.E. connectors	2		E				
S.E. connector	1		E				
Meter base	1		E				
#2-3 S.E. cable	25 ft.		C Ft.				
Miscellaneous	Lot						
TOTAL MATERIAL COST							
TOTAL LABOR COST - GROUP 1	7.00 WH		Hr				
GROUP 2	10.00 WH		Hr				
GROUP 3	13.00 WH		Hr				
GROUP 4	16.00 WH		Hr				
DIRECT JOB EXPENSE							
TOTAL PRIME COST				$	$	$	$

Service-Entrance, 150A, Copper S.E. Cable

No. 1 Wire

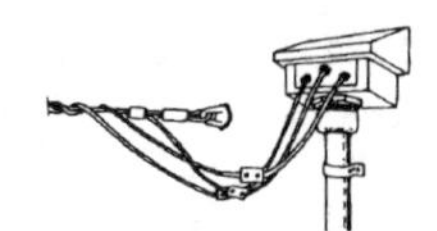

Cost Item	Quantity	Price or Rate	Per	Installation Groups 1	2	3	4
MATERIAL							
Service head	1		E				
Raintight S.E. connectors	2		E				
S.E. connector	1		E				
Meter base	1		E				
#1-3 S.E. cable	25 ft.		C Ft.				
Miscellaneous	Lot						
TOTAL MATERIAL COST							
TOTAL LABOR COST - GROUP 1	8.00 WH		Hr				
GROUP 2	11.00 WH		Hr				
GROUP 3	14.00 WH		Hr				
GROUP 4	16.50 WH		Hr				
DIRECT JOB EXPENSE							
TOTAL PRIME COST				$	$	$	$

Service-Entrance, 200A, Copper S.E. Cable

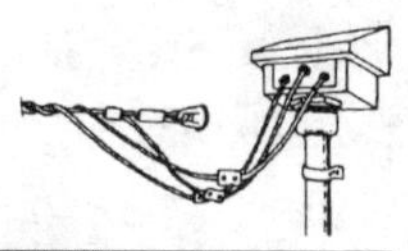

No. 2/0 Wire

Cost Item	Quantity	Price or Rate	Per	Installation Groups			
				1	2	3	4
MATERIAL							
Service head	1		E				
Raintight S.E. connectors	2		E				
S.E. connector	1		E				
Meter base	1		E				
#2/0-3 S.E. cable	25 ft.		C Ft.				
Miscellaneous	Lot						
TOTAL MATERIAL COST							
TOTAL LABOR COST - GROUP 1	9.00 WH		Hr				
GROUP 2	12.00 WH		Hr				
GROUP 3	15.00 WH		Hr				
GROUP 4	18.00 WH		Hr				
DIRECT JOB EXPENSE							
TOTAL PRIME COST				$	$	$	$

Conduit Systems

A conduit system consists of an electrical wiring system in which one or more individual conductors are pulled into a conduit or similar housing, usually after the raceway system has been completely installed. The basic raceways or conduit systems for residential services include:

- Rigid steel conduit
- Electrical metallic tubing (EMT)
- Intermediate metal conduit
- Rigid nonmetallic conduit

Other raceways systems are permitted for residential services, but the ones listed above are the most popular.

Conduits are available in standardized sizes and serve primarily to provide mechanical protection for the wires run inside, and, in the case of metallic raceways, to provide a continuously grounded system. Metallic raceways, properly installed, provide the greatest degree of mechanical and grounding protection and provide maximum protection against fire hazards for the electrical system. However, they are the most expensive to install.

Environmental conditions in certain areas of the United States necessitate local ordinances to stipulate the type of service entrances permitted. Consequently, some areas surpass the *NEC* in their requirements. Always consult the electrical inspection office in the jurisdiction in which the service is to be installed before finalizing the estimate for a residential service.

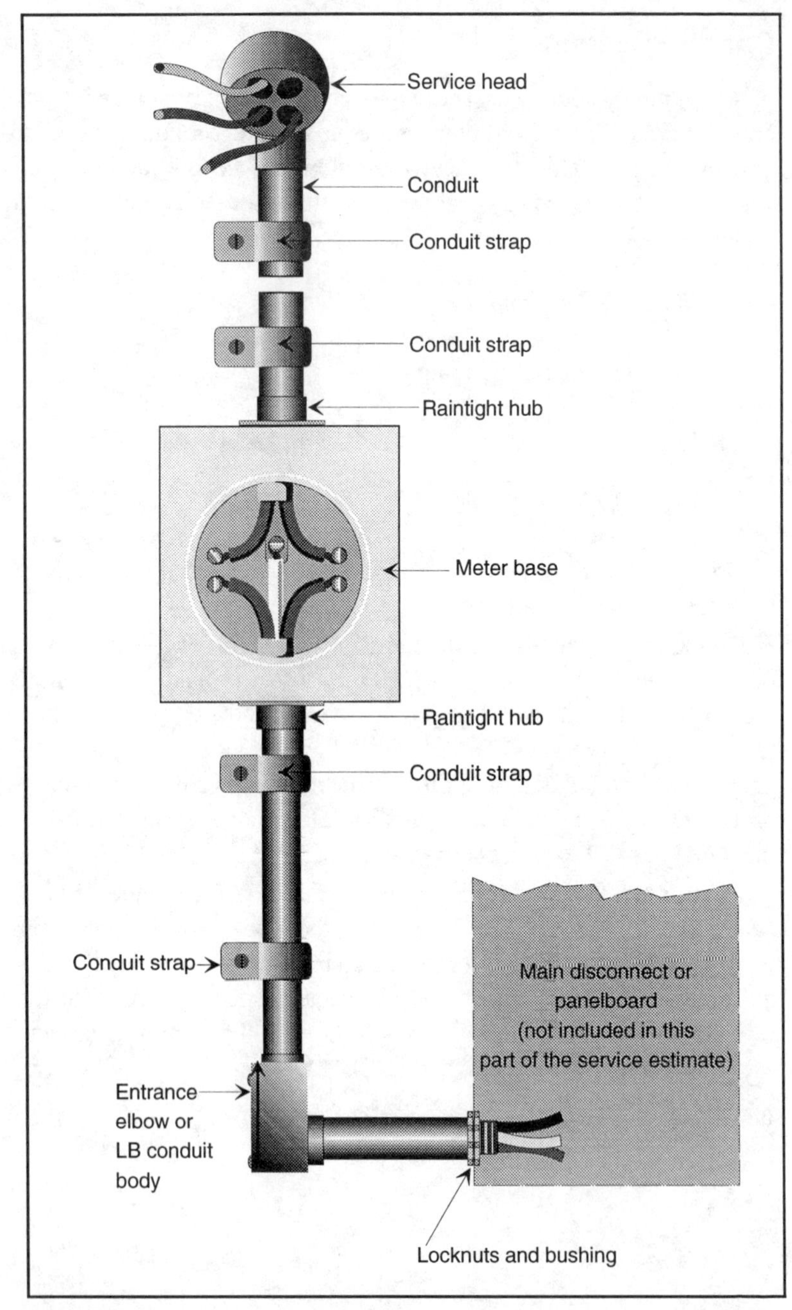

Figure 5-5: Components of a rigid conduit system used to contain service-entrance conductors.

Service-Entrance, 100A, Rigid Steel Conduit with Aluminum Conductors

No. 2 Wire

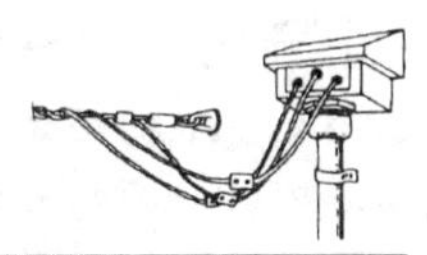

Cost Item	Quantity	Price or Rate	Per	Installation Groups 1	2	3	4
MATERIAL							
Service head	1		E				
Entrance elbow or LB conduit body	1		E				
Meter base with raintight conduit hubs	1		E				
1¼" rigid steel conduit	25 ft.		M Ft.				
#2 THHN conductors	90 ft.		C Ft.				
Miscellaneous	Lot						
TOTAL MATERIAL COST							
TOTAL LABOR COST - GROUP 1	8.00 WH		Hr				
GROUP 2	12.00 WH		Hr				
GROUP 3	15.00 WH		Hr				
GROUP 4	19.00 WH		Hr				
DIRECT JOB EXPENSE							
TOTAL PRIME COST				$	$	$	$

Service-Entrance, 125A, Rigid Steel Conduit with Aluminum Conductors

No. 1/0 Wire

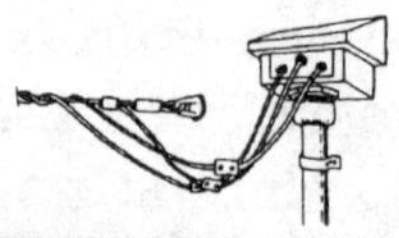

Cost Item	Quantity	Price or Rate	Per	Installation Groups			
				1	2	3	4
MATERIAL							
Service head	1		E				
Entrance elbow or LB conduit body	1		E				
Meter base with raintight conduit hubs	1		E				
1¼" rigid steel conduit	25 ft.		M Ft.				
#1/0 THHN conductors	90 ft.		C Ft.				
Miscellaneous	Lot						
TOTAL MATERIAL COST							
TOTAL LABOR COST - GROUP 1	9.00 WH		Hr				
GROUP 2	13.00 WH		Hr				
GROUP 3	16.00 WH		Hr				
GROUP 4	20.00 WH		Hr				
DIRECT JOB EXPENSE							
TOTAL PRIME COST				$	$	$	$

Service-Entrance, 150A, Rigid Steel Conduit with Aluminum Conductors

No. 2/0 Wire

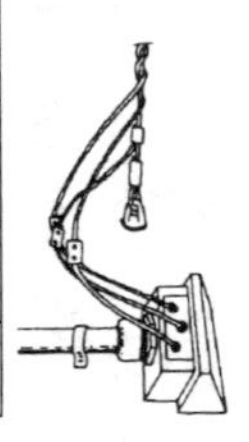

Cost Item	Quantity	Price or Rate	Per	Installation Groups			
MATERIAL				1	2	3	4
Service head	1		E				
Entrance elbow or LB conduit body	1		E				
Meter base with raintight conduit hubs	1		E				
1½" rigid steel conduit	25 ft.		M Ft.				
#2/0 THHN conductors	90 ft.		C Ft.				
Miscellaneous	Lot						
TOTAL MATERIAL COST							
TOTAL LABOR COST - GROUP 1	10.00 WH		Hr				
GROUP 2	15.00 WH		Hr				
GROUP 3	19.00 WH		Hr				
GROUP 4	24.00 WH		Hr				
DIRECT JOB EXPENSE							
TOTAL PRIME COST				$	$	$	$

Service-Entrance, 200A, Rigid Steel Conduit with Aluminum Conductors

No. 4/0 Wire

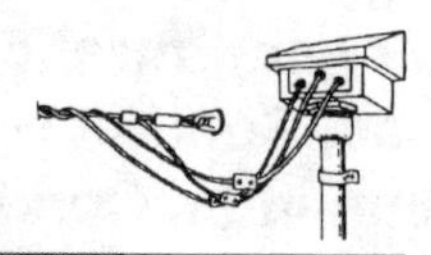

Cost Item	Quantity	Price or Rate	Per	Installation Groups 1	2	3	4
MATERIAL							
Service head	1		E				
Entrance elbow or LB conduit body	1		E				
Meter base with raintight conduit hubs	1		E				
2½" rigid steel conduit	25 ft.		M Ft.				
#4/0 THHN conductors	90 ft.		C Ft.				
Miscellaneous	Lot						
TOTAL MATERIAL COST							
TOTAL LABOR COST - GROUP 1	11.00 WH		Hr				
GROUP 2	17.00 WH		Hr				
GROUP 3	23.00 WH		Hr				
GROUP 4	29.00 WH		Hr				
DIRECT JOB EXPENSE							
TOTAL PRIME COST				$	$	$	$

Service-Entrance, 100A, Rigid Steel Conduit with Copper Conductors

No. 4 Wire

Cost Item	Quantity	Price or Rate	Per	Installation Groups 1	2	3	4
MATERIAL							
Service head	1		E				
Entrance elbow or LB conduit body	1		E				
Meter base with raintight conduit hubs	1		E				
1¼" rigid steel conduit	25 ft.		C Ft.				
#4 THHN conductors	90 ft.		M Ft.				
Miscellaneous	Lot						
TOTAL MATERIAL COST							
TOTAL LABOR COST - GROUP 1	8.00 WH		Hr				
GROUP 2	12.00 WH		Hr				
GROUP 3	15.00 WH		Hr				
GROUP 4	19.00 WH		Hr				
DIRECT JOB EXPENSE							
TOTAL PRIME COST				$	$	$	$

Service-Entrance, 125A, Rigid Steel Conduit with Copper Conductors

No. 2 Wire

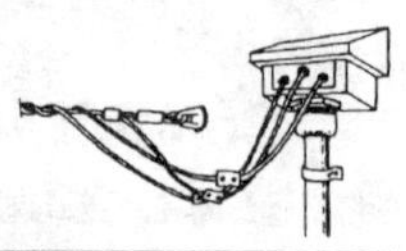

Cost Item	Quantity	Price or Rate	Per	Installation Groups 1	2	3	4
MATERIAL							
Service head	1		E				
Entrance elbow or LB conduit body	1		E				
Meter base with raintight conduit hubs	1		E				
1¼" rigid steel conduit	25 ft.		M Ft.				
#2 THHN conductors	90 ft.		C Ft.				
Miscellaneous	Lot						
TOTAL MATERIAL COST							
TOTAL LABOR COST - GROUP 1	9.00 WH		Hr				
GROUP 2	13.00 WH		Hr				
GROUP 3	16.00 WH		Hr				
GROUP 4	20.00 WH		Hr				
DIRECT JOB EXPENSE							
TOTAL PRIME COST				$	$	$	$

Service-Entrance, 150A, Rigid Steel Conduit with Copper Conductors

No. 1 Wire

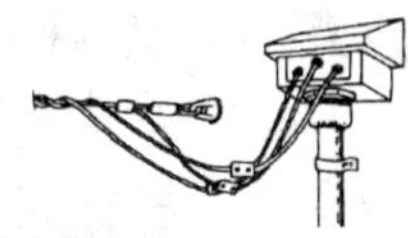

Cost Item	Quantity	Price or Rate	Per	Installation Groups			
				1	2	3	4
MATERIAL							
Service head	1		E				
Entrance elbow or LB conduit body	1		E				
Meter base with raintight conduit hubs	1		E				
1½" rigid steel conduit	25 ft.		M Ft.				
#2/0 THHN conductors	90 ft.		C Ft.				
Miscellaneous	Lot						
TOTAL MATERIAL COST							
TOTAL LABOR COST - GROUP 1	10.00 WH		Hr				
GROUP 2	15.00 WH		Hr				
GROUP 3	19.00 WH		Hr				
GROUP 4	24.00 WH		Hr				
DIRECT JOB EXPENSE							
TOTAL PRIME COST				$	$	$	$

Service-Entrance, 200A, Rigid Steel Conduit with Copper Conductors

No. 2/0 Wire

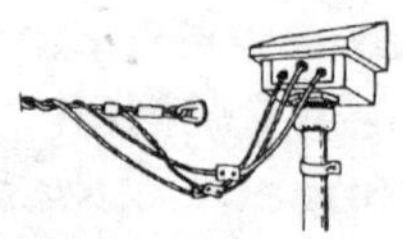

Cost Item	Quantity	Price or Rate	Per	Installation Groups 1	2	3	4
MATERIAL							
Service head	1		E				
Entrance elbow or LB conduit body	1		E				
Meter base with raintight conduit hubs	1		E				
1½" rigid steel conduit	25 ft.		M Ft.				
#2/0 THHN conductors	90 ft.		C Ft.				
Miscellaneous	Lot						
TOTAL MATERIAL COST							
TOTAL LABOR COST - GROUP 1	11.00 WH		Hr				
GROUP 2	17.00 WH		Hr				
GROUP 3	23.00 WH		Hr				
GROUP 4	29.00 WH		Hr				
DIRECT JOB EXPENSE							
TOTAL PRIME COST				$	$	$	$

Service-Entrance, 100A, Rigid PVC Conduit with Aluminum Conductors

No. 2 Wire

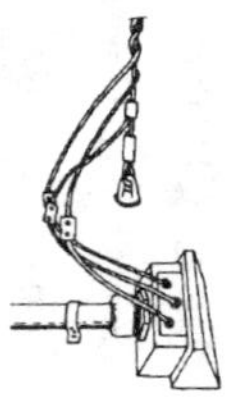

Cost Item	Quantity	Price or Rate	Per	Installation Groups			
MATERIAL				1	2	3	4
Service head	1		E				
Entrance elbow or LB conduit body	1		E				
Meter base with raintight conduit hubs	1		E				
1¼" rigid PVC conduit	25 ft.		C Ft.				
#2 THHN conductors	90 ft.		M Ft.				
Miscellaneous	Lot						
TOTAL MATERIAL COST							
TOTAL LABOR COST - GROUP 1	7.00 WH		Hr				
GROUP 2	10.00 WH		Hr				
GROUP 3	14.00 WH		Hr				
GROUP 4	17.00 WH		Hr				
DIRECT JOB EXPENSE							
TOTAL PRIME COST				$	$	$	$

Service-Entrance, 125A, Rigid PVC Conduit with Aluminum Conductors

No. 1/0 Wire

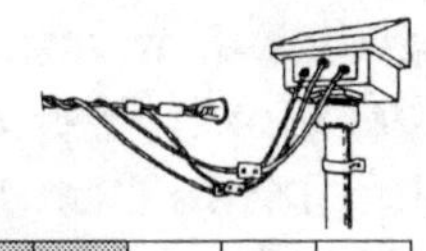

Cost Item	Quantity	Price or Rate	Per	Installation Groups			
				1	2	3	4
MATERIAL							
Service head	1		E				
Entrance elbow or LB conduit body	1		E				
Meter base with raintight conduit hubs	1		E				
1¼" rigid PVC conduit	25 ft.		C Ft.				
#1/0 THHN conductors	90 ft.		C Ft.				
Miscellaneous	Lot						
TOTAL MATERIAL COST							
TOTAL LABOR COST - GROUP 1	8.00 WH		Hr				
GROUP 2	11.00 WH		Hr				
GROUP 3	15.00 WH		Hr				
GROUP 4	18.00 WH		Hr				
DIRECT JOB EXPENSE							
TOTAL PRIME COST				$	$	$	$

Service-Entrance, 150A, Rigid PVC Conduit with Aluminum Conductors

No. 2/0 Wire

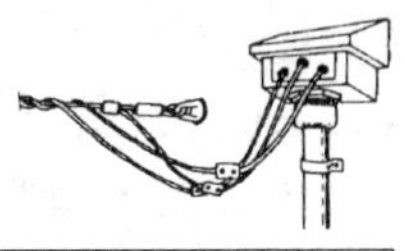

Cost Item	Quantity	Price or Rate	Per	Installation Groups			
				1	2	3	4
MATERIAL							
Service head	1		E				
Entrance elbow or LB conduit body	1		E				
Meter base with raintight conduit hubs	1		E				
1½" rigid PVC conduit	25 ft.		C Ft.				
#2/0 THHN conductors	90 ft.		M Ft.				
Miscellaneous	Lot						
TOTAL MATERIAL COST							
TOTAL LABOR COST - GROUP 1	9.00 WH		Hr				
GROUP 2	13.00 WH		Hr				
GROUP 3	18.00 WH		Hr				
GROUP 4	21.00 WH		Hr				
DIRECT JOB EXPENSE							
TOTAL PRIME COST				$	$	$	$

Service-Entrance, 200A, Rigid PVC Conduit with Aluminum Conductors

No. 4/0 Wire

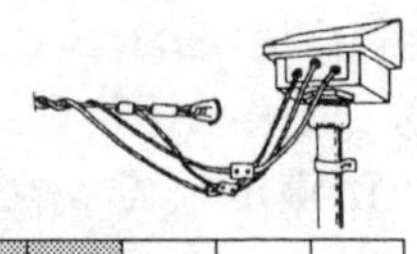

Cost Item	Quantity	Price or Rate	Per	Installation Groups 1	2	3	4
MATERIAL							
Service head	1		E				
Entrance elbow or LB conduit body	1		E				
Meter base with raintight conduit hubs	1		E				
2½" rigid steel conduit	25 ft.		C Ft.				
#4/0 THHN conductors	90 ft.		M Ft.				
Miscellaneous	Lot						
TOTAL MATERIAL COST							
TOTAL LABOR COST - GROUP 1	10.00 WH		Hr				
GROUP 2	14.00 WH		Hr				
GROUP 3	19.00 WH		Hr				
GROUP 4	23.00 WH		Hr				
DIRECT JOB EXPENSE							
TOTAL PRIME COST				$	$	$	$

Service Masts

Where a service mast is used for the support of service-drop conductors, the conduit — serving as the support — must be of adequate strength to safely withstand the strain imposed by the service drop. If the conduit size is not sufficient to withstand the strain, braces or guys may be used to support the mast.

When conduit is used for the support, all conduit fittings must be identified for use with the service mast. Such fittings include the following:

- Service head
- Service-drop support
- Roof flange
- Conduit seal

The through-the-roof conduit must be of sufficient height so that the service conductors — when attached to the mast — will not be less than 18 inches above the overhang portion of the roof. This minimum height (18 inches) is permitted provided the mast protrudes through the roof no more than 4 feet from the edge.

In general, the installation of a through-the-roof service mast requires boring a hole through the roof and also through the soffit below with a hole saw to allow the conduit to pass from the meter base to a sufficient height above the roof. A metal flange, containing a conduit seal, is then inserted over the protruding conduit and sealed in place to prevent moisture from entering the building structure.

Both materials and worker-hours increase over a conventional conduit service when dealing with a service mast. These increases are detailed in the charts to follow.

A "mast kit," as indicated in the estimating charts, consists of a service head, guy/conductor terminator, roof flange, and conduit seal. Such kits are normally supplied with these items for service masts and are readily available from electrical suppliers.

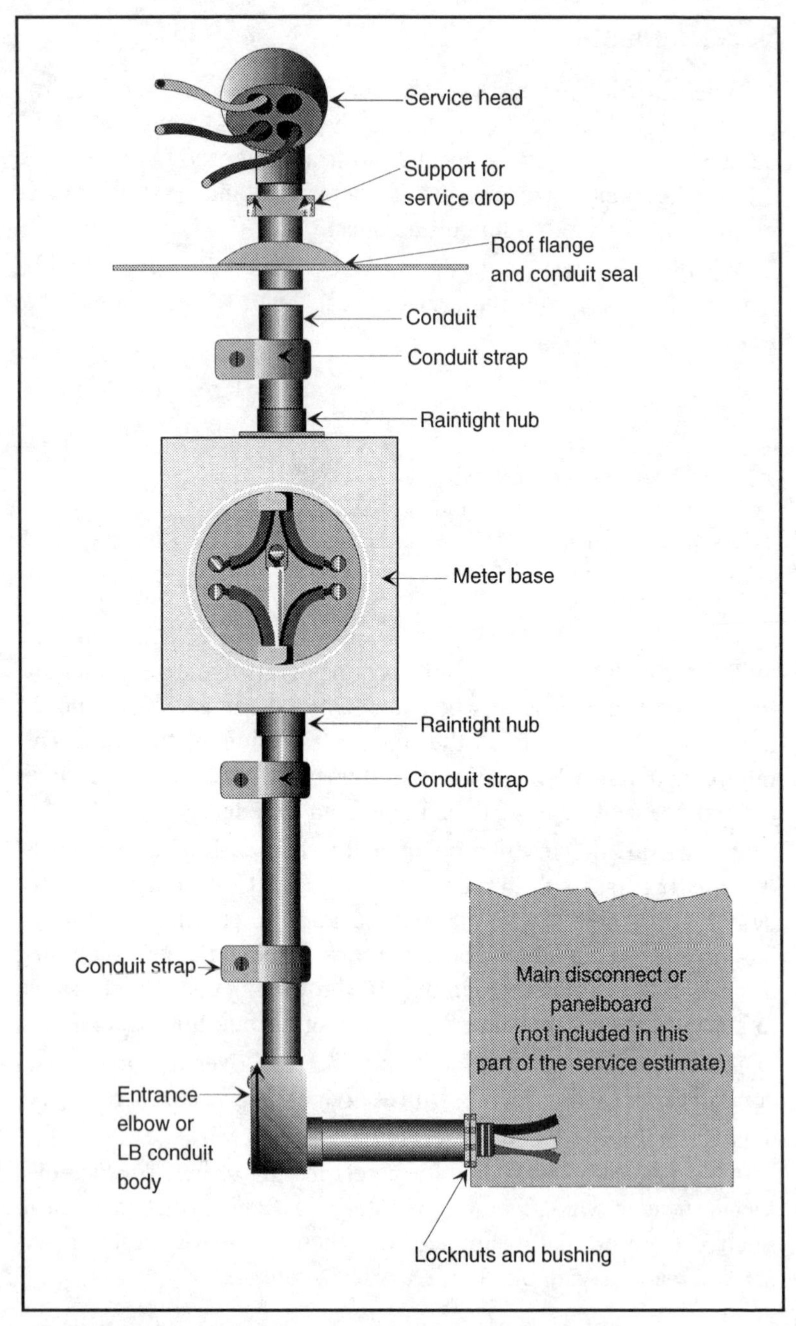

Figure 5-6: Components of a residential service mast.

Service-Entrance, 100A, Mast Type with Aluminum Conductors

No. 2 Wire

Cost Item	Quantity	Price or Rate	Per	Installation Groups 1	2	3	4
MATERIAL							
Service mast kit	1		E				
Entrance elbow or LB conduit body	1		E				
Meter base with raintight conduit hubs	1		E				
2½" rigid steel conduit	25 ft.		C Ft.				
#2 THHN conductors	90 ft.		M Ft.				
Miscellaneous	Lot						
TOTAL MATERIAL COST							
TOTAL LABOR COST - GROUP 1	7.90 WH		Hr				
GROUP 2	12.10 WH		Hr				
GROUP 3	16.50 WH		Hr				
GROUP 4	21.30 WH		Hr				
DIRECT JOB EXPENSE							
TOTAL PRIME COST				$	$	$	$

Service-Entrance, 125A, Mast Type with Aluminum Conductors

No. 1/0 Wire

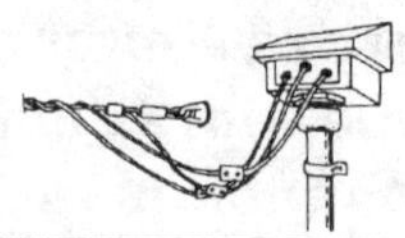

Cost Item	Quantity	Price or Rate	Per	Installation Groups 1	2	3	4
MATERIAL							
Service mast kit	1		E				
Entrance elbow or LB conduit body	1		E				
Meter base with raintight conduit hubs	1		E				
2½" rigid steel conduit	25 ft.		C Ft.				
#1/0 THHN conductors	90 ft.		M Ft.				
Miscellaneous	Lot						
TOTAL MATERIAL COST							
TOTAL LABOR COST - GROUP 1	9.00 WH		Hr				
GROUP 2	12.10 WH		Hr				
GROUP 3	16.50 WH		Hr				
GROUP 4	21.15 WH		Hr				
DIRECT JOB EXPENSE							
TOTAL PRIME COST				$	$	$	$

Service-Entrance, 150A, Mast Type with Aluminum Conductors

No. 2/0 Wire

Cost Item	Quantity	Price or Rate	Per	Installation Groups			
				1	2	3	4
MATERIAL							
Service mast kit	1		E				
Entrance elbow or LB conduit body	1		E				
Meter base with raintight conduit hubs	1		E				
2½" rigid steel conduit	25 ft.		C Ft.				
#2/0 THHN conductors	90 ft.		M Ft.				
Miscellaneous	Lot						
TOTAL MATERIAL COST							
TOTAL LABOR COST - GROUP 1	12.10 WH		Hr				
GROUP 2	18.10 WH		Hr				
GROUP 3	24.50 WH		Hr				
GROUP 4	31.30 WH		Hr				
DIRECT JOB EXPENSE							
TOTAL PRIME COST				$	$	$	$

Service-Entrance, 200A, Mast Type with Aluminum Conductors

No. 4/0 Wire

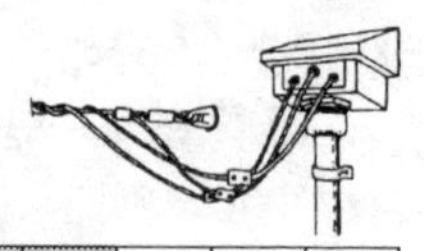

Cost Item	Quantity	Price or Rate	Per	Installation Groups 1	2	3	4
MATERIAL							
Service mast kit	1		E				
Entrance elbow or LB conduit body	1		E				
Meter base with raintight conduit hubs	1		E				
2½" rigid steel conduit	25 ft.		C Ft.				
#4/0 THHN conductors	90 ft.		M Ft.				
Miscellaneous	Lot						
TOTAL MATERIAL COST							
TOTAL LABOR COST - GROUP 1	11.90 WH		Hr				
GROUP 2	18.10 WH		Hr				
GROUP 3	24.50 WH		Hr				
GROUP 4	31.30 WH		Hr				
DIRECT JOB EXPENSE							
TOTAL PRIME COST				$	$	$	$

Service-Entrance, 100A, Mast Type with Copper Conductors

No. 4 Wire

Cost Item	Quantity	Price or Rate	Per	Installation Groups			
MATERIAL				1	2	3	4
Service mast kit	1		E				
Entrance elbow or LB conduit body	1		E				
Meter base with raintight conduit hubs	1		E				
2½" rigid steel conduit	25 ft.		C Ft.				
#4 THHN conductors	90 ft.		M Ft.				
Miscellaneous	Lot						
TOTAL MATERIAL COST							
TOTAL LABOR COST - GROUP 1	8.00 WH		Hr				
GROUP 2	12.20 WH		Hr				
GROUP 3	16.75 WH		Hr				
GROUP 4	21.50 WH		Hr				
DIRECT JOB EXPENSE							
TOTAL PRIME COST				$	$	$	$

Service-Entrance, 125A, Mast Type with Copper Conductors

No. 2 Wire

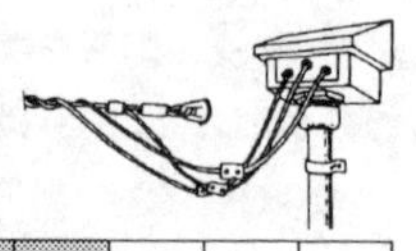

Cost Item	Quantity	Price or Rate	Per	Installation Groups			
MATERIAL				1	2	3	4
Service mast kit	1		E				
Entrance elbow or LB conduit body	1		E				
Meter base with raintight conduit hubs	1		E				
2½" rigid steel conduit	25 ft.		C Ft.				
#2 THHN conductors	90 ft.		M Ft.				
Miscellaneous	Lot						
TOTAL MATERIAL COST							
TOTAL LABOR COST - GROUP 1	9.00 WH		Hr				
GROUP 2	12.00 WH		Hr				
GROUP 3	17.00 WH		Hr				
GROUP 4	21.00 WH		Hr				
DIRECT JOB EXPENSE							
TOTAL PRIME COST				$	$	$	$

Service-Entrance, 150A, Mast Type with Copper Conductors

No. 1 Wire

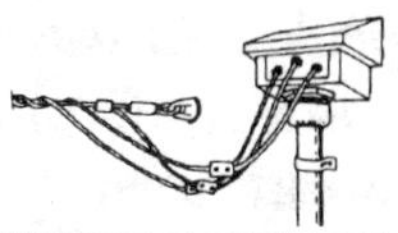

Cost Item	Quantity	Price or Rate	Per	Installation Groups 1	2	3	4
MATERIAL							
Service mast kit	1		E				
Entrance elbow or LB conduit body	1		E				
Meter base with raintight conduit hubs	1		E				
2½" rigid steel conduit	25 ft.		C Ft.				
#1 THHN conductors	90 ft.		M Ft.				
Miscellaneous	Lot						
TOTAL MATERIAL COST							
TOTAL LABOR COST - GROUP 1	12.00 WH		Hr				
GROUP 2	18.00 WH		Hr				
GROUP 3	24.00 WH		Hr				
GROUP 4	31.00 WH		Hr				
DIRECT JOB EXPENSE							
TOTAL PRIME COST				$	$	$	$

Service-Entrance, 200A, Mast Type with Copper Conductors

No. 2/0 Wire

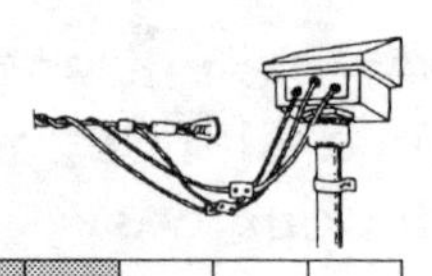

Cost Item	Quantity	Price or Rate	Per	Installation Groups			
				1	2	3	4
MATERIAL							
Service mast kit	1		E				
Entrance elbow or LB conduit body	1		E				
Meter base with raintight conduit hubs	1		E				
2½" rigid steel conduit	25 ft.		C Ft.				
#2/0 THHN conductors	90 ft.		M Ft.				
Miscellaneous	Lot						
TOTAL MATERIAL COST							
TOTAL LABOR COST - GROUP 1	13.00 WH		Hr				
GROUP 2	19.00 WH		Hr				
GROUP 3	26.00 WH		Hr				
GROUP 4	33.00 WH		Hr				
DIRECT JOB EXPENSE							
TOTAL PRIME COST				$	$	$	$

Underground Service Laterals

When the service conductors to a building or property are routed underground, these conductors are known as the service lateral, defined as follows:

- *Service lateral:* The underground conductors through which service is supplied between the power company's distribution facilities and the first point of their connection to the building or area service facilities.

The exact requirements of the electrical contractor's responsibility for underground service laterals vary from jurisdiction to jurisdiction. In some areas, the electrical contractor is required to open a trench from the power pole or pad-mounted transformer, mount the meter base, and install an empty conduit from the meter base to a minimum depth of 18 inches below grade. The power company will then install the service lateral from their transformer to the meter base.

In other cases, the power company will install all materials from their power supply to, and including, the meter base. The electrical contractor is then responsible for installing the service-conductor run from the meter to the service disconnecting means.

The estimating charts to follow list material and corresponding labor to install the meter base, a length of empty rigid conduit from the meter base to 18 inches below grade, and rigid conduit from the meter base (including conductors) to the main disconnect cabinet. Any variations in these specifications should be handled prior to reaching a selling price for bidding work.

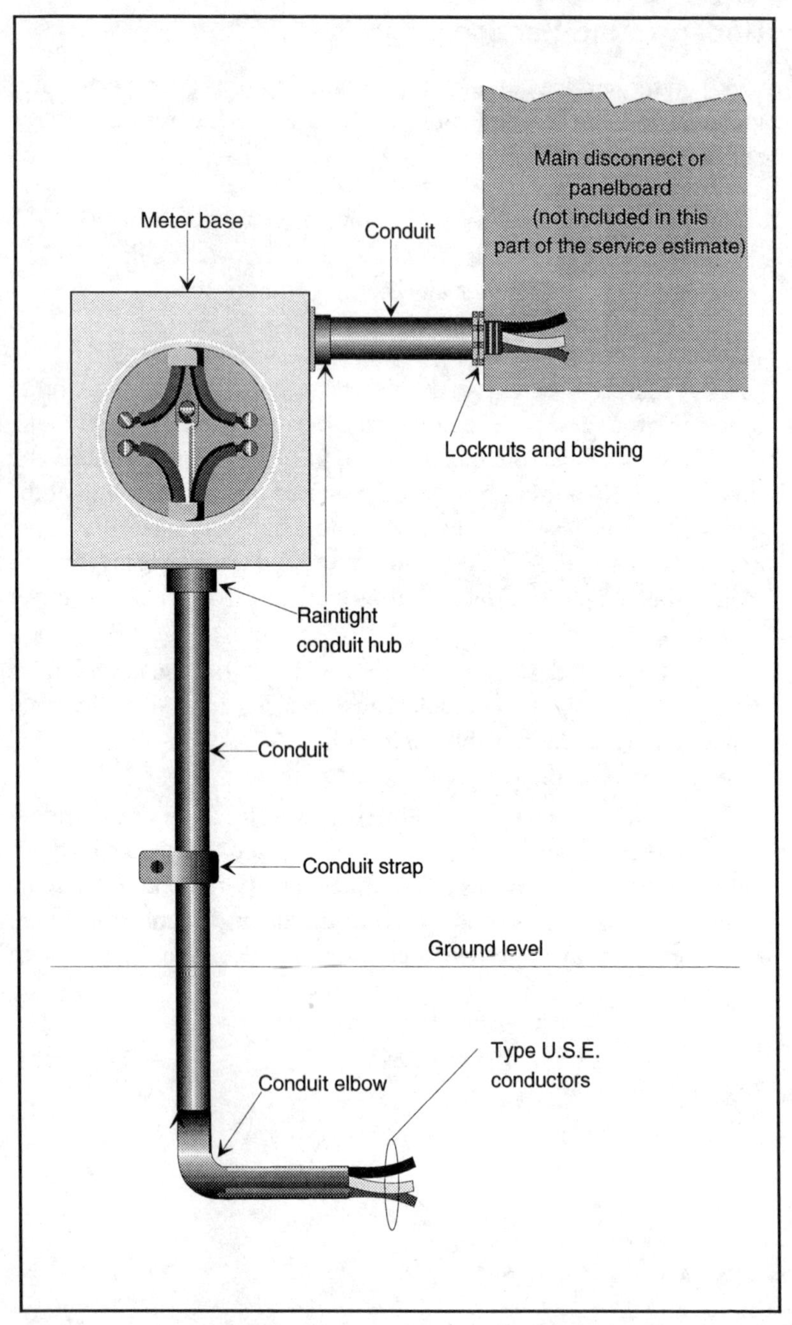

Figure 5-7: Components of an underground service lateral.

Service Lateral, 100A Copper Conductors

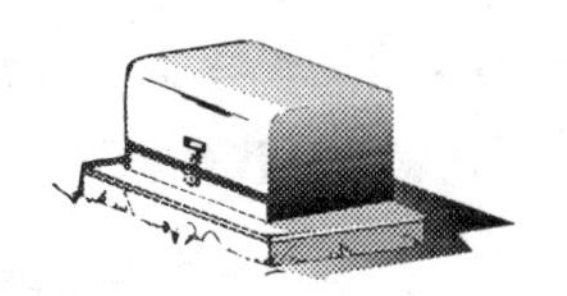

No. 4 Wire

Cost Item	Quantity	Price or Rate	Per	Installation Groups 1	2	3	4
MATERIAL							
Meter base with raintight conduit hubs	1		E				
1¼" rigid steel conduit	10 ft.		C Ft.				
1¼" rigid steel conduit 90° elbow							
#4 THHN conductors	16 ft.		M Ft.				
Miscellaneous	Lot						
TOTAL MATERIAL COST							
TOTAL LABOR COST - GROUP 1	5.90 WH		Hr				
GROUP 2	10.10 WH		Hr				
GROUP 3	14.50 WH		Hr				
GROUP 4	19.30 WH		Hr				
DIRECT JOB EXPENSE							
TOTAL PRIME COST				$	$	$	$

Service Lateral, 125A
Copper Conductors

No. 2 Wire

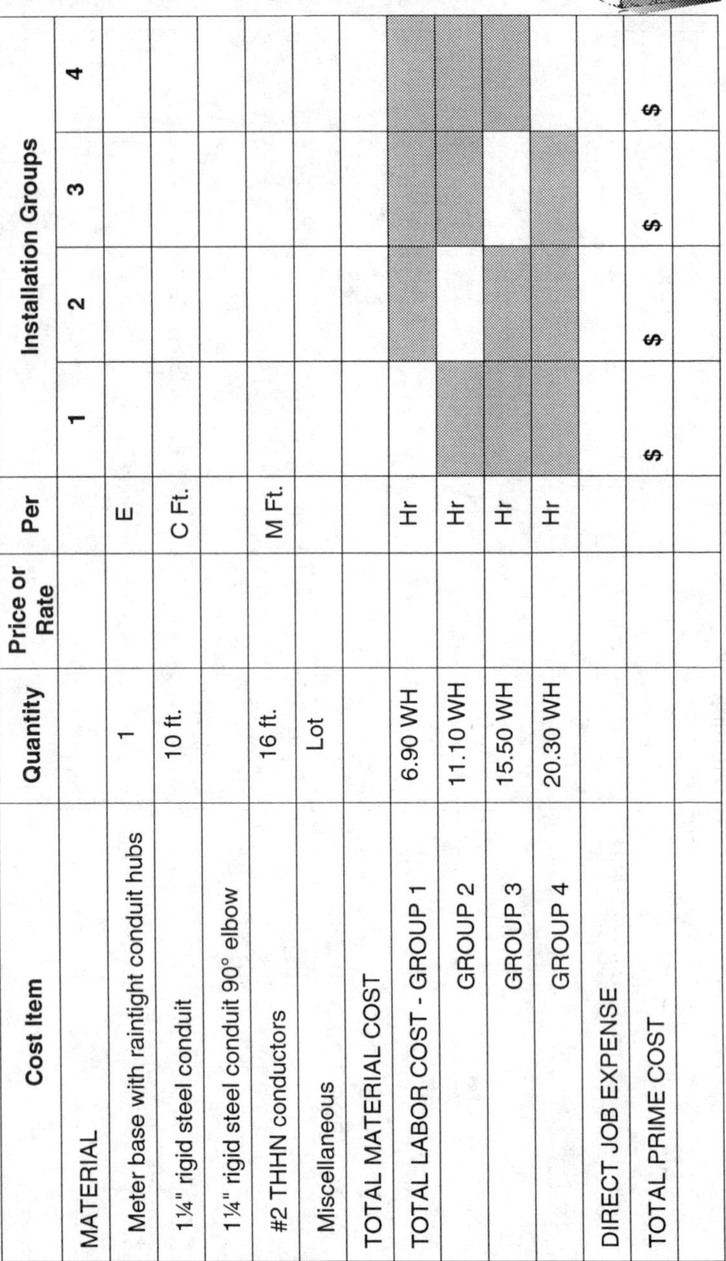

Cost Item	Quantity	Price or Rate	Per	Installation Groups			
MATERIAL				1	2	3	4
Meter base with raintight conduit hubs	1		E				
1¼" rigid steel conduit	10 ft.		C Ft.				
1¼" rigid steel conduit 90° elbow							
#2 THHN conductors	16 ft.		M Ft.				
Miscellaneous	Lot						
TOTAL MATERIAL COST							
TOTAL LABOR COST - GROUP 1	6.90 WH		Hr				
GROUP 2	11.10 WH		Hr				
GROUP 3	15.50 WH		Hr				
GROUP 4	20.30 WH		Hr				
DIRECT JOB EXPENSE							
TOTAL PRIME COST				$	$	$	$

Service Lateral, 150A Copper Conductors

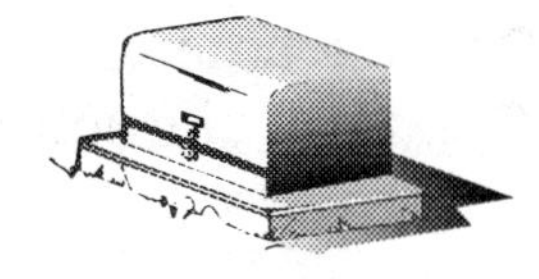

No. 1 Wire

Cost Item	Quantity	Price or Rate	Per	Installation Groups 1	2	3	4
MATERIAL							
Meter base with raintight conduit hubs	1		E				
1¼" rigid steel conduit	10 ft.		C Ft.				
1¼" rigid steel conduit 90° elbow							
#1 THHN conductors	16 ft.		M Ft.				
Miscellaneous	Lot						
TOTAL MATERIAL COST							
TOTAL LABOR COST - GROUP 1	7.90 WH		Hr				
GROUP 2	12.10 WH		Hr				
GROUP 3	16.50 WH		Hr				
GROUP 4	21.30 WH		Hr				
DIRECT JOB EXPENSE							
TOTAL PRIME COST				$	$	$	$

Service Lateral, 200A
Copper Conductors

No. 2/0 Wire

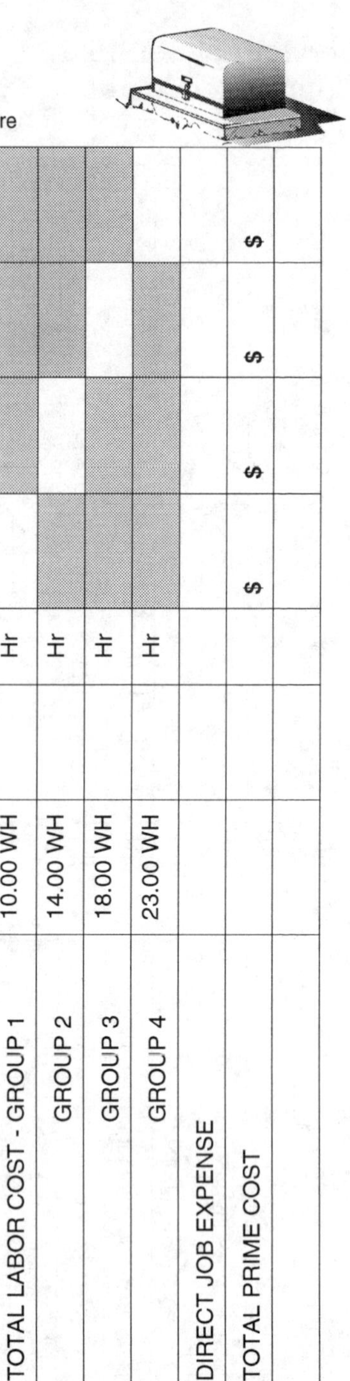

Cost Item	Quantity	Price or Rate	Per	Installation Groups			
				1	2	3	4
MATERIAL							
Meter base with raintight conduit hubs	1		E				
1½" rigid steel conduit	10 ft.		C Ft.				
1½" rigid steel conduit 90° elbow							
#2/0 THHN conductors	16 ft.		M Ft.				
Miscellaneous	Lot						
TOTAL MATERIAL COST							
TOTAL LABOR COST - GROUP 1	10.00 WH		Hr				
GROUP 2	14.00 WH		Hr				
GROUP 3	18.00 WH		Hr				
GROUP 4	23.00 WH		Hr				
DIRECT JOB EXPENSE							
TOTAL PRIME COST				$	$	$	$

Service Equipment

The *NEC* defines "Service Equipment" as follows:

> "The necessary equipment, usually consisting of a circuit breaker or switch and fuses, and their accessories, located near the point of entrance of supply conductors to a building or other structure, or an otherwise defined area, and intended to constitute the main control and means of cutoff of the supply."

The *NEC* requires that a service be provided with a disconnecting means for all conductors in the building or structure from the service-entrance conductors. This disconnecting means should be located at or near the point where the service conductors enter the building. Furthermore, the disconnecting means must be located at a readily accessible point, either inside or outside the building. Adequate access and working space must be provided all around the disconnecting means.

Overcurrent protection is required both at the main source and for all individual feeders and branch circuits to protect the electrical installation against ground faults and overloads.

The majority of overcurrent devices are housed in some type of enclosure. The National Electrical Manufacturers Association (NEMA) has established enclosure designations because individually enclosed overcurrent protective devices are used in so many different types of locations. These designations are described in the table in Figure 5-8 on the next page.

Enclosure	Explanation
NEMA Type 1 General Purpose	To prevent accidental contact with enclosed apparatus. Suitable for application indoors where not exposed to unusual service conditions.
NEMA Type 3 Weatherproof (Weather Resistant)	Protection against specified weather hazards. Suitable for use oudoors
NEMA Type 3R Raintight	Protects against entrance of water from a rain. Suitable general outdoor application not requiring sleetproof.
NEMA Type 4 Watertight	Designed to exclude water applied in form of hose stream. To protect against stream of water during cleaning operations, etc.
NEMA Type 5 Dusttight	Constructed so that dust will not enter the enclosed case. Being replaced in some equipment by NEMA 12 types.
NEMA Type 7	Designed to meet application requirements of the *National Electrical Code* for Class I, Hazardous locations (explosive atmospheres). Circuit interruption occurs in air.
NEMA Type 9	Designed to meet application requirements of *National Electrical Code* for Class II Hazardous Locations (combustible dusts, etc.)
NEMA Type 12	For use in areas where it is desired to exclude dust, lint, fibers and filings, or oil or coolant seepage.

Figure 5-8: NEMA classifications of safety switches.

Fusible Safety Switches, 30A - 200A
NEMA Type 1

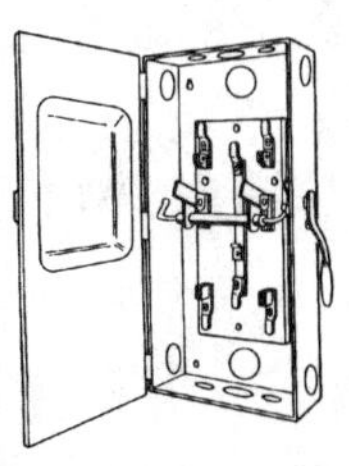

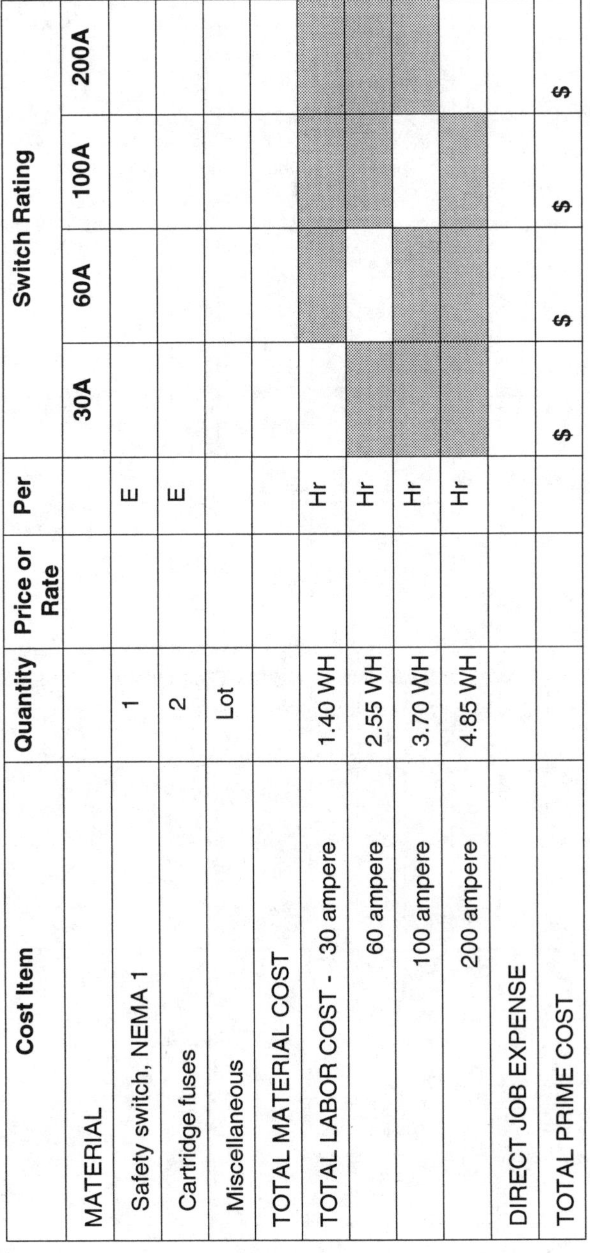

Cost Item	Quantity	Price or Rate	Per	Switch Rating			
				30A	60A	100A	200A
MATERIAL							
Safety switch, NEMA 1	1		E				
Cartridge fuses	2		E				
Miscellaneous	Lot						
TOTAL MATERIAL COST							
TOTAL LABOR COST - 30 ampere	1.40 WH		Hr				
60 ampere	2.55 WH		Hr				
100 ampere	3.70 WH		Hr				
200 ampere	4.85 WH		Hr				
DIRECT JOB EXPENSE							
TOTAL PRIME COST				$	$	$	$

Fusible Safety Switches, 30A - 200A
Raintight (Weatherproof)

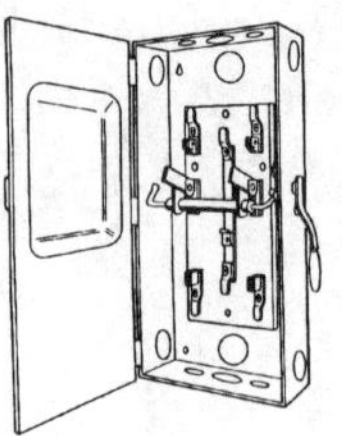

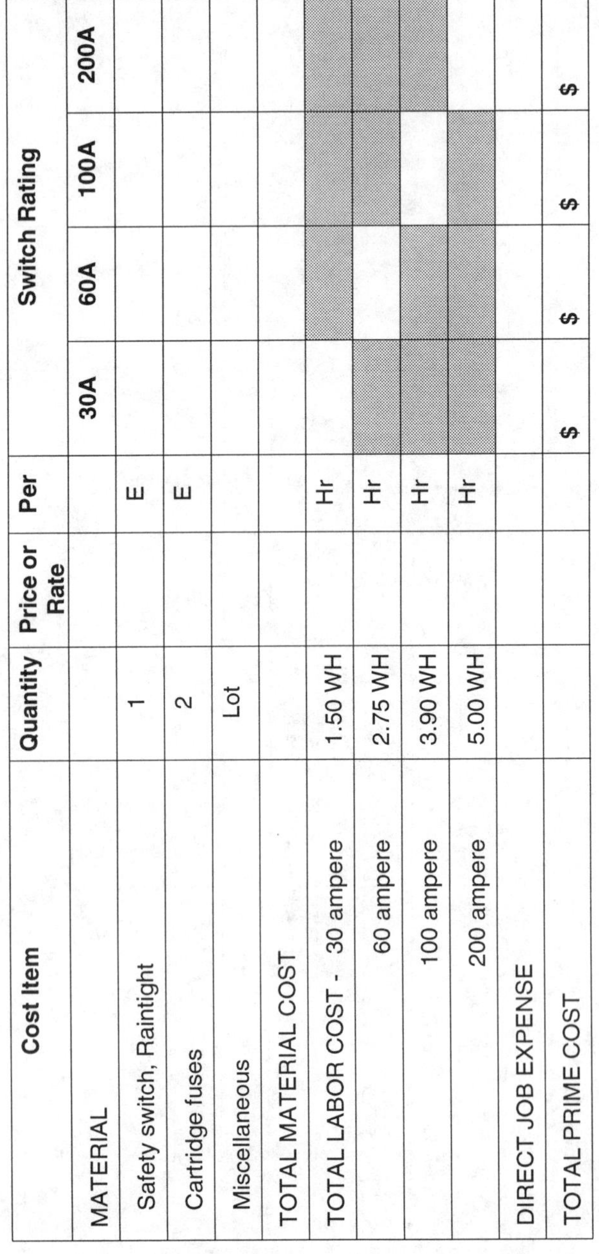

Cost Item	Quantity	Price or Rate	Per	Switch Rating			
				30A	60A	100A	200A
MATERIAL							
Safety switch, Raintight	1		E				
Cartridge fuses	2		E				
Miscellaneous	Lot						
TOTAL MATERIAL COST							
TOTAL LABOR COST - 30 ampere	1.50 WH		Hr				
60 ampere	2.75 WH		Hr				
100 ampere	3.90 WH		Hr				
200 ampere	5.00 WH		Hr				
DIRECT JOB EXPENSE							
TOTAL PRIME COST				$	$	$	$

Nonfusible Safety Switches, 30A - 200A
NEMA Type 1

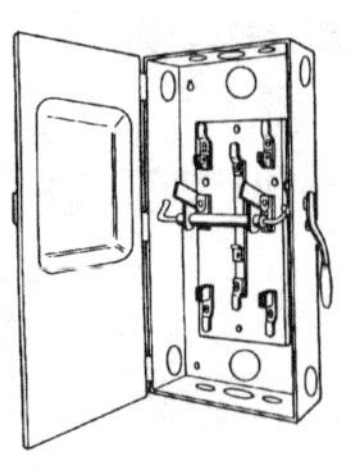

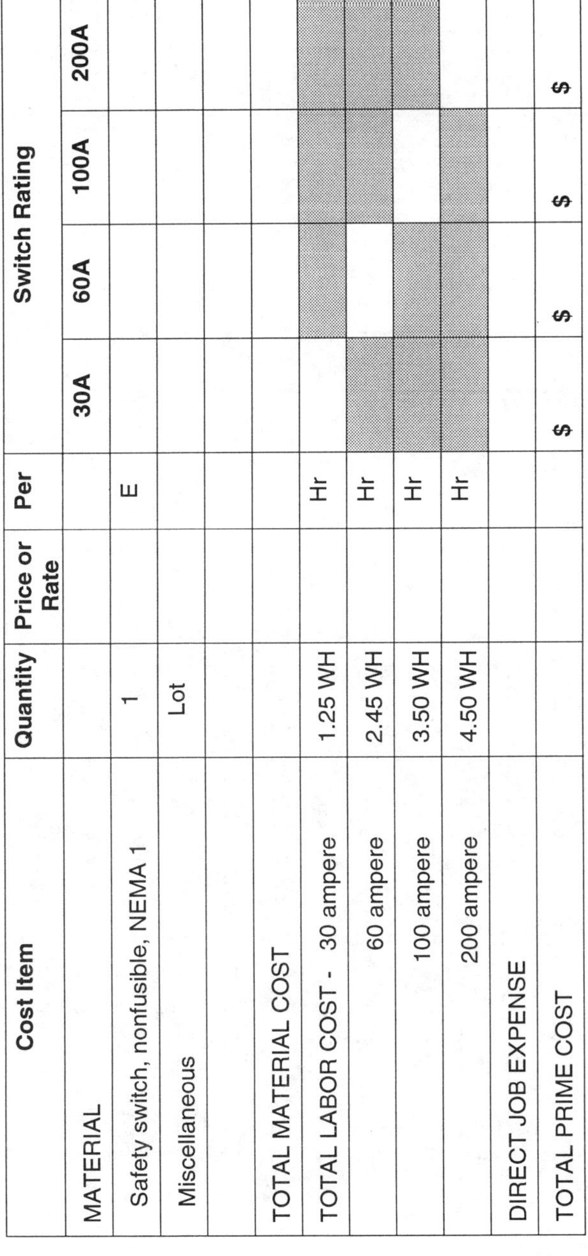

Cost Item	Quantity	Price or Rate	Per	Switch Rating			
				30A	60A	100A	200A
MATERIAL							
Safety switch, nonfusible, NEMA 1	1		E				
Miscellaneous	Lot						
TOTAL MATERIAL COST							
TOTAL LABOR COST - 30 ampere	1.25 WH		Hr				
60 ampere	2.45 WH		Hr				
100 ampere	3.50 WH		Hr				
200 ampere	4.50 WH		Hr				
DIRECT JOB EXPENSE							
TOTAL PRIME COST				$	$	$	$

Nonfusible Safety Switches, 30A - 200A Raintight (Weatherproof)

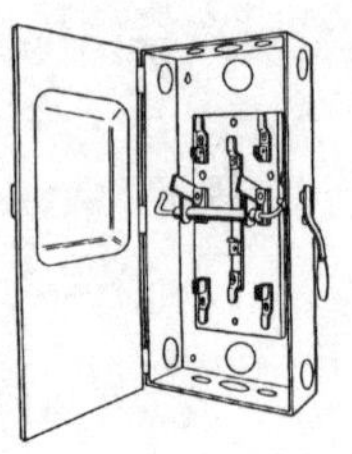

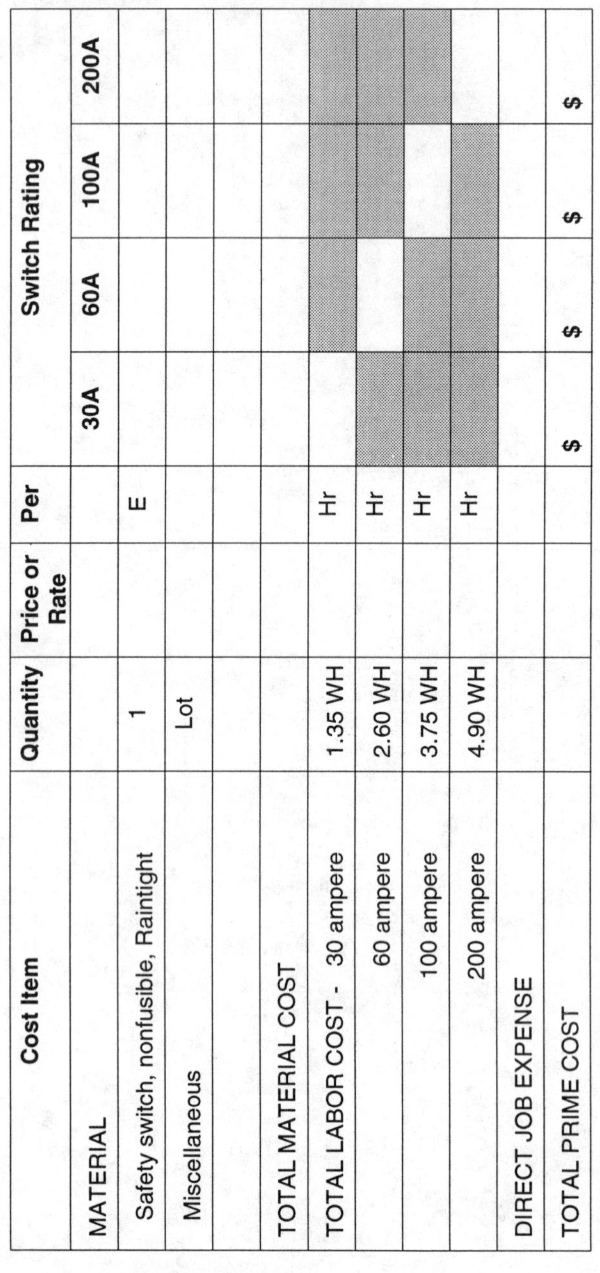

Cost Item	Quantity	Price or Rate	Per	Switch Rating			
				30A	60A	100A	200A
MATERIAL							
Safety switch, nonfusible, Raintight	1		E				
Miscellaneous	Lot						
TOTAL MATERIAL COST							
TOTAL LABOR COST - 30 ampere	1.35 WH		Hr				
60 ampere	2.60 WH		Hr				
100 ampere	3.75 WH		Hr				
200 ampere	4.90 WH		Hr				
DIRECT JOB EXPENSE							
TOTAL PRIME COST				$	$	$	$

Circuit Breaker Load Center, 100A Main Breaker

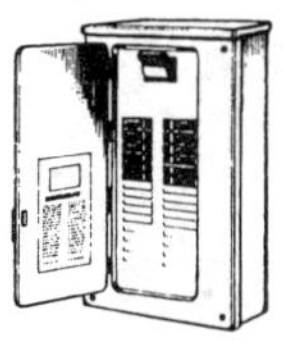

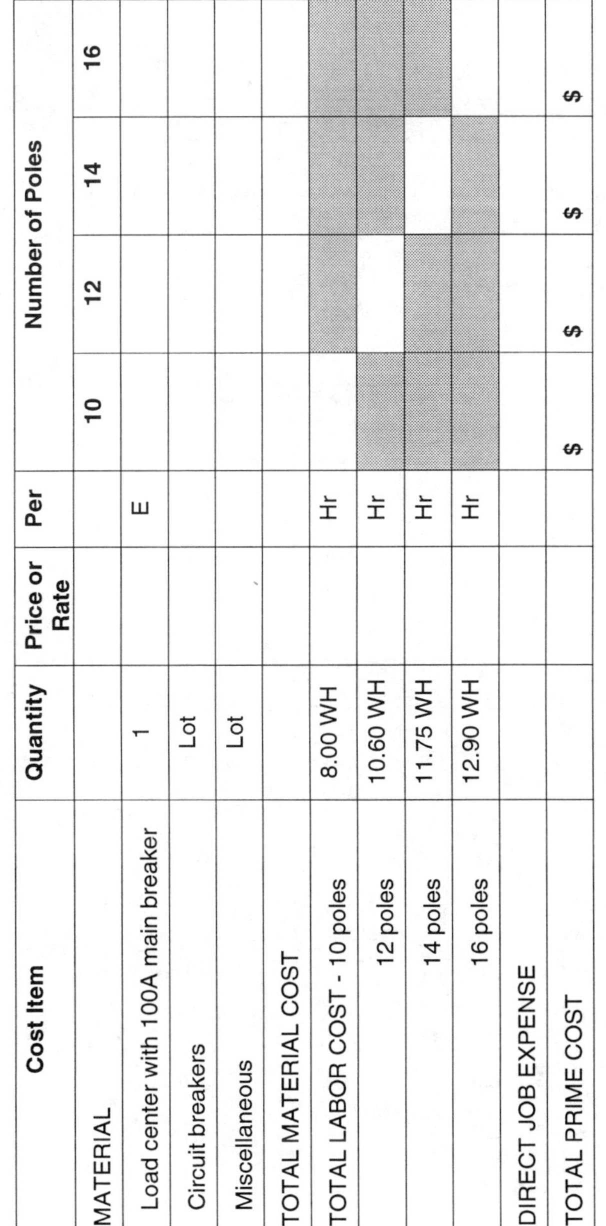

Cost Item	Quantity	Price or Rate	Per	Number of Poles			
				10	12	14	16
MATERIAL							
Load center with 100A main breaker	1		E				
Circuit breakers	Lot						
Miscellaneous	Lot						
TOTAL MATERIAL COST							
TOTAL LABOR COST - 10 poles	8.00 WH		Hr				
12 poles	10.60 WH		Hr				
14 poles	11.75 WH		Hr				
16 poles	12.90 WH		Hr				
DIRECT JOB EXPENSE							
TOTAL PRIME COST				$	$	$	$

Circuit Breaker Load Center, 150A Main Breaker

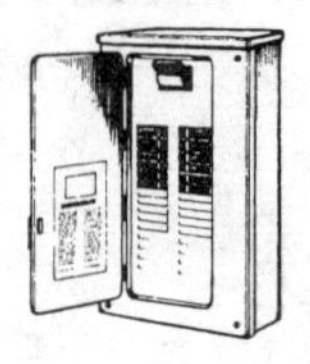

Cost Item	Quantity	Price or Rate	Per	Number of Poles			
				16	18	20	24
MATERIAL							
Load center with 150A main breaker	1		E				
Circuit breakers	Lot						
Miscellaneous	Lot						
TOTAL MATERIAL COST							
TOTAL LABOR COST - 16 poles	13.00 WH		Hr				
18 poles	14.60 WH		Hr				
20 poles	15.75 WH		Hr				
24 poles	18.90 WH		Hr				
DIRECT JOB EXPENSE							
TOTAL PRIME COST				$	$	$	$

Circuit Breaker Load Center, 200A Main Breaker

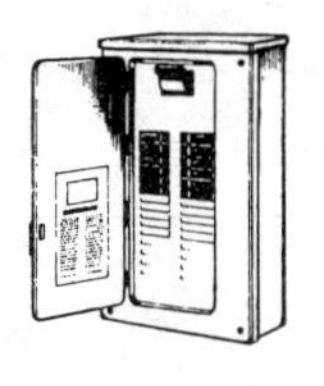

Cost Item	Quantity	Price or Rate	Per	Number of Poles			
				24	32	36	40
MATERIAL							
Load center with 200A main breaker	1		E				
Circuit breakers	Lot						
Miscellaneous	Lot						
TOTAL MATERIAL COST							
TOTAL LABOR COST - 24 poles	19.00 WH		Hr				
32 poles	21.60 WH		Hr				
36 poles	24.00 WH		Hr				
40 poles	26.90 WH		Hr				
DIRECT JOB EXPENSE							
TOTAL PRIME COST				$	$	$	$

Circuit Breaker Load Center, 100A Main Lugs Only

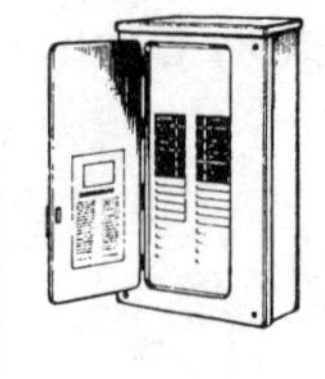

Cost Item	Quantity	Price or Rate	Per	Number of Poles			
				10	12	14	16
MATERIAL							
Load center with 100A main lugs only	1		E				
Circuit breakers	Lot						
Miscellaneous	Lot						
TOTAL MATERIAL COST							
TOTAL LABOR COST - 10 poles	7.00 WH		Hr				
12 poles	9.00 WH		Hr				
14 poles	10.00 WH		Hr				
16 poles	11.00 WH		Hr				
DIRECT JOB EXPENSE							
TOTAL PRIME COST				$	$	$	$

Circuit Breaker Load Center, 150A Main Lugs Only

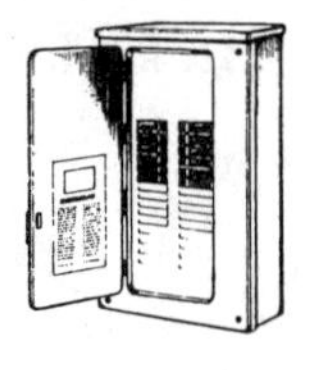

Cost Item	Quantity	Price or Rate	Per	Number of Poles			
				16	18	20	24
MATERIAL							
Load center with 150A main lugs only	1		E				
Circuit breakers	Lot						
Miscellaneous	Lot						
TOTAL MATERIAL COST							
TOTAL LABOR COST - 16 poles	12.00 WH		Hr				
18 poles	13.00 WH		Hr				
20 poles	14.00 WH		Hr				
24 poles	17.00 WH		Hr				
DIRECT JOB EXPENSE							
TOTAL PRIME COST				$	$	$	$

Circuit Breaker Load Center, 200A Main Lugs Only

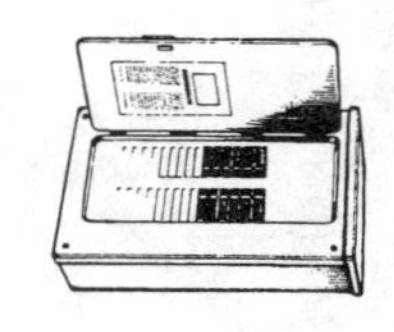

Cost Item	Quantity	Price or Rate	Per	Number of Poles 24	32	36	40
MATERIAL							
Load center with 200A main lugs only	1		E				
Circuit breakers	Lot						
Miscellaneous	Lot						
TOTAL MATERIAL COST							
TOTAL LABOR COST - 24 poles	18.00 WH		Hr				
32 poles	20.00 WH		Hr				
36 poles	23.00 WH		Hr				
40 poles	25.00 WH		Hr				
DIRECT JOB EXPENSE							
TOTAL PRIME COST				$	$	$	$

Grounding

Methods of grounding an electric service are covered in *NEC* Section 250-81. In general, all of the following (if available) and any made electrodes must be bonded together to form the grounding electrode system.

- An underground metallic water pipe in direct contact with the earth for no less than 10 feet.

- The metal frame of a building where effectively grounded.

- An electrode encased by at least two inches of concrete, located within and near the bottom of a concrete foundation or footing that is in direct contact with the earth. Furthermore, this electrode must be at least 20 feet long and must be made of electrically conductive coated steel reinforcing bars or rods of not less than ½-inch diameter, or consisting of at least 20 feet of bare copper conductor not smaller than No. 2 AWG wire size.

- A ground ring encircling the building or structure, in direct contact with the earth at a depth below grade not less than 2½ feet. This ring must consist of at least 20 feet of bare copper conductor not smaller than No. 2 AWG wire size.

In most residential structures, only the water pipe will be available, and this water pipe must be supplemented by an additional electrode as specified in *NEC* Sections 250-81(a) and 250-83. In most cases, the supplemental electrode will consist of either a driven rod or pipe electrode. See Figure 5-9 on the next page. The estimating tables to follow include provisions for correct grounding on the majority of residential occupancies.

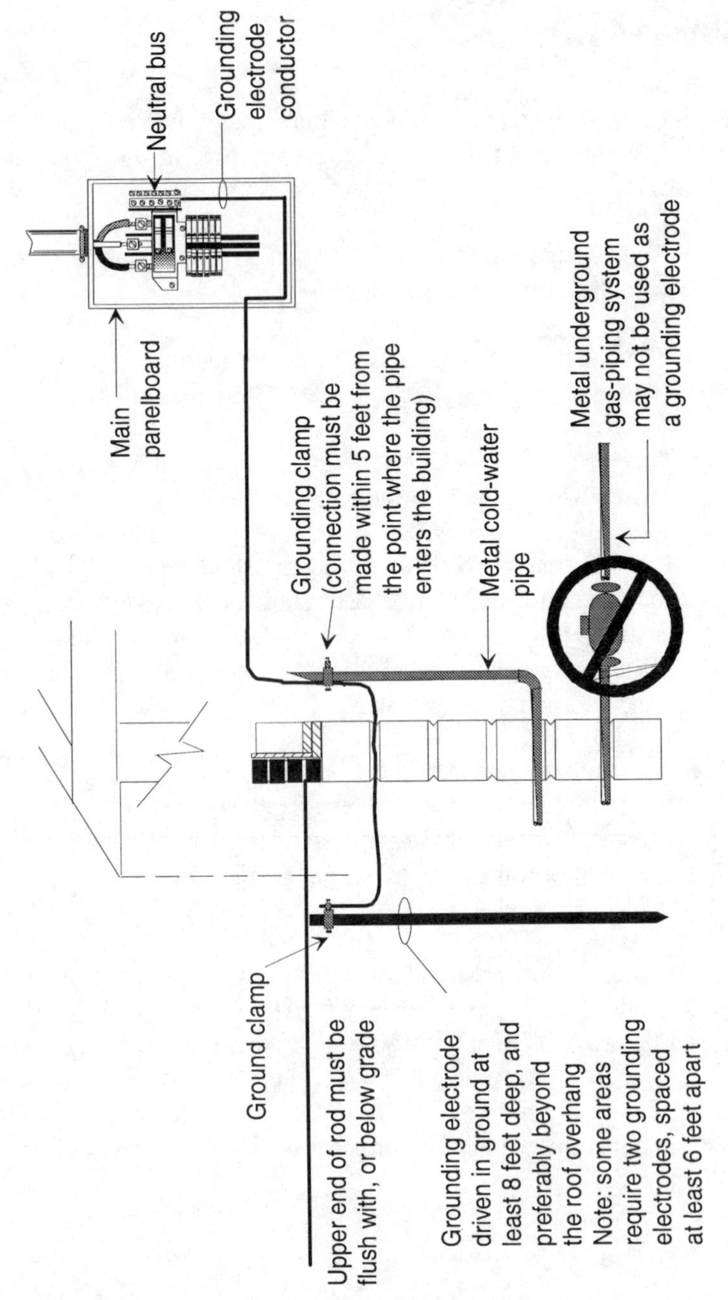

Figure 5-9: Components of a residential service grounding system.

Service Grounding - 100A to 200A

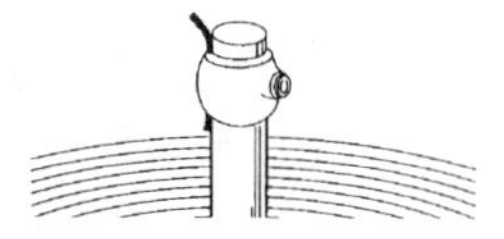

Cost Item	Quantity	Price or Rate	Per	Switch Rating			
				100A	125A	150A	200A
MATERIAL							
Grounding clamps	2		E				
Grounding wire, bare copper	25 ft.		M Ft.				
Copperweld grounding rod	1		E				
Miscellaneous	Lot						
TOTAL MATERIAL COST							
TOTAL LABOR COST - 100 ampere	1.35 WH		Hr				
125 ampere	1.60 WH		Hr				
150 ampere	1.75 WH		Hr				
200 ampere	1.90 WH		Hr				
DIRECT JOB EXPENSE							
TOTAL PRIME COST				$	$	$	$

Selling-Price Tables

The selling-price tables to follow are designed for quick unit pricing of electric services, using all of the popular wiring methods. Before arriving at a selling price, the preceding tables must be completed (filled in with appropriate prices). Then the prime cost determined for each wiring situation in the preceding tables should be entered in the corresponding selling-price tables to follow. Once the prime cost has been entered in each selling-price table, overhead and profit factors are calculated to arrive at a total unit selling price as described in detail in Chapter 2.

100A Aluminum S.E. Cable	No. 2 wire	GROUPS			
		1	2	3	4
Prime Cost		$	$	$	$
Overhead	_% of prime cost				
Total Cost		$	$	$	$
Profit	_% of total cost				
Selling Price		$	$	$	$

125A Aluminum S.E. Cable	No. 1/0 wire	GROUPS			
		1	2	3	4
Prime Cost		$	$	$	$
Overhead	_% of prime cost				
Total Cost		$	$	$	$
Profit	_% of total cost				
Selling Price		$	$	$	$

150A Aluminum S.E. Cable	No. 2/0 wire	GROUPS			
		1	2	3	4
Prime Cost		$	$	$	$
Overhead	_% of prime cost				
Total Cost		$	$	$	$
Profit	_% of total cost				
Selling Price		$	$	$	$

200A Aluminum S.E. Cable	No. 4/0 wire	GROUPS			
		1	2	3	4
Prime Cost		$	$	$	$
Overhead	_% of prime cost				
Total Cost		$	$	$	$
Profit	_% of total cost				
Selling Price		$	$	$	$

100A Copper S.E. Cable	No. 4 wire	GROUPS			
		1	2	3	4
Prime Cost		$	$	$	$
Overhead	_% of prime cost				
Total Cost		$	$	$	$
Profit	_% of total cost				
Selling Price		$	$	$	$

125A Copper S.E. Cable	No. 2 wire	GROUPS			
		1	2	3	4
Prime Cost		$	$	$	$
Overhead	_% of prime cost				
Total Cost		$	$	$	$
Profit	_% of total cost				
Selling Price		$	$	$	$

150A Copper S.E. Cable	No. 1 wire	GROUPS			
		1	2	3	4
Prime Cost		$	$	$	$
Overhead	_% of prime cost				
Total Cost		$	$	$	$
Profit	_% of total cost				
Selling Price		$	$	$	$

200A Copper S.E. Cable	No. 2/0 wire	GROUPS			
		1	2	3	4
Prime Cost		$	$	$	$
Overhead	_% of prime cost				
Total Cost		$	$	$	$
Profit	_% of total cost				
Selling Price		$	$	$	$

100A Aluminum Rigid Conduit	No. 2 wire	GROUPS			
		1	2	3	4
Prime Cost		$	$	$	$
Overhead	_% of prime cost				
Total Cost		$	$	$	$
Profit	_% of total cost				
Selling Price		$	$	$	$

125A Aluminum Rigid Conduit	No. 1/0 wire	GROUPS			
		1	2	3	4
Prime Cost		$	$	$	$
Overhead	_% of prime cost				
Total Cost		$	$	$	$
Profit	_% of total cost				
Selling Price		$	$	$	$

150A Aluminum Rigid Conduit	No. 2/0 wire	GROUPS			
		1	2	3	4
Prime Cost		$	$	$	$
Overhead	_% of prime cost				
Total Cost		$	$	$	$
Profit	_% of total cost				
Selling Price		$	$	$	$

200A Aluminum Rigid Conduit	No. 4/0 wire	GROUPS			
		1	2	3	4
Prime Cost		$	$	$	$
Overhead	_% of prime cost				
Total Cost		$	$	$	$
Profit	_% of total cost				
Selling Price		$	$	$	$

100A Copper Rigid Conduit	No. 4 wire	GROUPS			
		1	2	3	4
Prime Cost		$	$	$	$
Overhead	_% of prime cost				
Total Cost		$	$	$	$
Profit	_% of total cost				
Selling Price		$	$	$	$

125A Copper Rigid Conduit	No. 2 wire	GROUPS			
		1	2	3	4
Prime Cost		$	$	$	$
Overhead	_% of prime cost				
Total Cost		$	$	$	$
Profit	_% of total cost				
Selling Price		$	$	$	$

150A Copper Rigid Conduit	No. 1 wire	GROUPS			
		1	2	3	4
Prime Cost		$	$	$	$
Overhead	_% of prime cost				
Total Cost		$	$	$	$
Profit	_% of total cost				
Selling Price		$	$	$	$

200A Copper Rigid Conduit	No. 2/0 wire	GROUPS			
		1	2	3	4
Prime Cost		$	$	$	$
Overhead	_% of prime cost				
Total Cost		$	$	$	$
Profit	_% of total cost				
Selling Price		$	$	$	$

100A Aluminum PVC Conduit	No. 2 wire	GROUPS			
		1	2	3	4
Prime Cost		$	$	$	$
Overhead	_% of prime cost				
Total Cost		$	$	$	$
Profit	_% of total cost				
Selling Price		$	$	$	$

125A Aluminum PVC Conduit	No. 1/0 wire	GROUPS			
		1	2	3	4
Prime Cost		$	$	$	$
Overhead	_% of prime cost				
Total Cost		$	$	$	$
Profit	_% of total cost				
Selling Price		$	$	$	$

150A Aluminum PVC Conduit	No. 2/0 wire	GROUPS			
		1	2	3	4
Prime Cost		$	$	$	$
Overhead	_% of prime cost				
Total Cost		$	$	$	$
Profit	_% of total cost				
Selling Price		$	$	$	$

200A Aluminum PVC Conduit	No. 4/0 wire	GROUPS			
		1	2	3	4
Prime Cost		$	$	$	$
Overhead	_% of prime cost				
Total Cost		$	$	$	$
Profit	_% of total cost				
Selling Price		$	$	$	$

100A Aluminum Service Mast	No. 2 wire	GROUPS			
		1	2	3	4
Prime Cost		$	$	$	$
Overhead	_% of prime cost				
Total Cost		$	$	$	$
Profit	_% of total cost				
Selling Price		$	$	$	$

125A Aluminum Service Mast	No. 1/0 wire	GROUPS			
		1	2	3	4
Prime Cost		$	$	$	$
Overhead	_% of prime cost				
Total Cost		$	$	$	$
Profit	_% of total cost				
Selling Price		$	$	$	$

150A Aluminum Service Mast	No. 2/0 wire	GROUPS			
		1	2	3	4
Prime Cost		$	$	$	$
Overhead	_% of prime cost				
Total Cost		$	$	$	$
Profit	_% of total cost				
Selling Price		$	$	$	$

200A Aluminum Service Mast	No. 4/0 wire	GROUPS			
		1	2	3	4
Prime Cost		$	$	$	$
Overhead	_% of prime cost				
Total Cost		$	$	$	$
Profit	_% of total cost				
Selling Price		$	$	$	$

100A Copper Service Mast	No. 4 wire	GROUPS			
		1	2	3	4
Prime Cost		$	$	$	$
Overhead	_% of prime cost				
Total Cost		$	$	$	$
Profit	_% of total cost				
Selling Price		$	$	$	$

125A Copper Service Mast	No. 2 wire	GROUPS			
		1	2	3	4
Prime Cost		$	$	$	$
Overhead	_% of prime cost				
Total Cost		$	$	$	$
Profit	_% of total cost				
Selling Price		$	$	$	$

150A Copper Service Mast	No. 1 wire	GROUPS			
		1	2	3	4
Prime Cost		$	$	$	$
Overhead	_% of prime cost				
Total Cost		$	$	$	$
Profit	_% of total cost				
Selling Price		$	$	$	$

200A Copper Service Mast	No. 2/0 wire	GROUPS			
		1	2	3	4
Prime Cost		$	$	$	$
Overhead	_% of prime cost				
Total Cost		$	$	$	$
Profit	_% of total cost				
Selling Price		$	$	$	$

100A Copper Service Lateral	No.4 wire	GROUPS			
		1	2	3	4
Prime Cost		$	$	$	$
Overhead	_% of prime cost				
Total Cost		$	$	$	$
Profit	_% of total cost				
Selling Price		$	$	$	$

125A Copper Service Lateral	No. 2 wire	GROUPS			
		1	2	3	4
Prime Cost		$	$	$	$
Overhead	_% of prime cost				
Total Cost		$	$	$	$
Profit	_% of total cost				
Selling Price		$	$	$	$

150A Copper Service Lateral	No. 1 wire	GROUPS			
		1	2	3	4
Prime Cost		$	$	$	$
Overhead	_% of prime cost				
Total Cost		$	$	$	$
Profit	_% of total cost				
Selling Price		$	$	$	$

200A Copper Service Lateral	No. 2/0 wire	GROUPS			
		1	2	3	4
Prime Cost		$	$	$	$
Overhead	_% of prime cost				
Total Cost		$	$	$	$
Profit	_% of total cost				
Selling Price		$	$	$	$

Fusible Safety Switch	NEMA 1	Switch Rating			
		30A	60A	100A	200A
Prime Cost		$	$	$	$
Overhead	_% of prime cost				
Total Cost		$	$	$	$
Profit	_% of total cost				
Selling Price		$	$	$	$

Fusible Safety Switch	Raintight (Weatherproof)	Switch Rating			
		30A	60A	100A	200A
Prime Cost		$	$	$	$
Overhead	_% of prime cost				
Total Cost		$	$	$	$
Profit	_% of total cost				
Selling Price		$	$	$	$

Nonfusible Safety Switch	NEMA 1	Switch Rating			
		30A	60A	100A	200A
Prime Cost		$	$	$	$
Overhead	_% of prime cost				
Total Cost		$	$	$	$
Profit	_% of total cost				
Selling Price		$	$	$	$

Nonfusible Safety Switch	Raintight (Weatherproof)	Switch Rating			
		30A	60A	100A	200A
Prime Cost		$	$	$	$
Overhead	_% of prime cost				
Total Cost		$	$	$	$
Profit	_% of total cost				
Selling Price		$	$	$	$

Circuit Breaker Load Center	100A Main Breaker	Number of Poles			
		10	12	14	16
Prime Cost		$	$	$	$
Overhead	_% of prime cost				
Total Cost		$	$	$	$
Profit	_% of total cost				
Selling Price		$	$	$	$

Circuit Breaker Load Center	150A Main Breaker	Number of Poles			
		16	18	20	24
Prime Cost		$	$	$	$
Overhead	_% of prime cost				
Total Cost		$	$	$	$
Profit	_% of total cost				
Selling Price		$	$	$	$

Circuit Breaker Load Center	200A Main Breaker	Number of Poles			
		24	32	36	40
Prime Cost		$	$	$	$
Overhead	_% of prime cost				
Total Cost		$	$	$	$
Profit	_% of total cost				
Selling Price		$	$	$	$

Circuit Breaker Load Center	100A Main Lugs Only	Number of Poles			
		10	12	14	16
Prime Cost		$	$	$	$
Overhead	_% of prime cost				
Total Cost		$	$	$	$
Profit	_% of total cost				
Selling Price		$	$	$	$

Circuit Breaker Load Center	150A Main Lugs Only	Number of Poles			
		16	18	20	24
Prime Cost		$	$	$	$
Overhead	_% of prime cost				
Total Cost		$	$	$	$
Profit	_% of total cost				
Selling Price		$	$	$	$

Circuit Breaker Load Center	200A Main Lugs Only	Number of Poles			
		24	32	36	40
Prime Cost		$	$	$	$
Overhead	_% of prime cost				
Total Cost		$	$	$	$
Profit	_% of total cost				
Selling Price		$	$	$	$

Service Grounding		Service Rating			
		100A	125A	150A	200A
Prime Cost		$	$	$	$
Overhead	_% of prime cost				
Total Cost		$	$	$	$
Profit	_% of total cost				
Selling Price		$	$	$	$

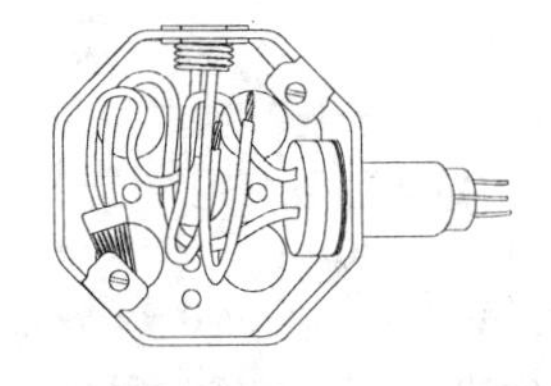

Chapter 6
Low-Voltage Wiring

One of the simplest and most common residential low-voltage systems is the residential door chime. Such systems usually contain a low-voltage source, one or more pushbuttons, low-voltage cable and a set of chimes. The quality of the chimes will range from a one-note device to those which "play" lengthy melodies.

The wiring diagram in Figure 6-1 shows a typical two-note chime controlled at two locations. One button, at the main entrance, will sound the two notes when pushed, while the other button, at the rear door, will sound only one note when pushed.

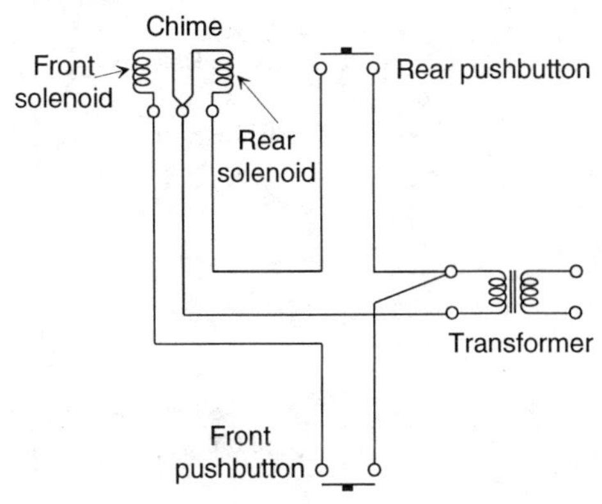

Figure 6-1: Wiring diagram of a two-note chime circuit.

In applications where lighting must be controlled from several points, or where there is a complexity of lighting or power circuits, or where flexibility is desirable in certain systems, low-voltage remote-controlled relay systems are frequently applied. These systems use special low-voltage components, operated from a transformer, to switch relays, which in turn control the standard line-voltage lighting circuits. Because the control wiring does not carry the direct line load, small lightweight cable can be used. It can be installed wherever and however convenient — placed behind moldings, stapled to woodwork, buried in shallow plaster channels, or installed in holes bored in wall studs.

Low-voltage switching is especially useful in rewiring existing buildings since the small cables are as easy to run as telephone wires. They are easy to hide behind baseboards or even behind quarter-round molding. The cable can be run exposed without being very noticeable because of its small size. A basic remote-control switching circuit is shown in Figure 6-2.

Furthermore, electrical contractors may be called upon to install low-voltage wiring and solid-state controls for HVAC and security/fire-alarm systems. See Figure 6-3.

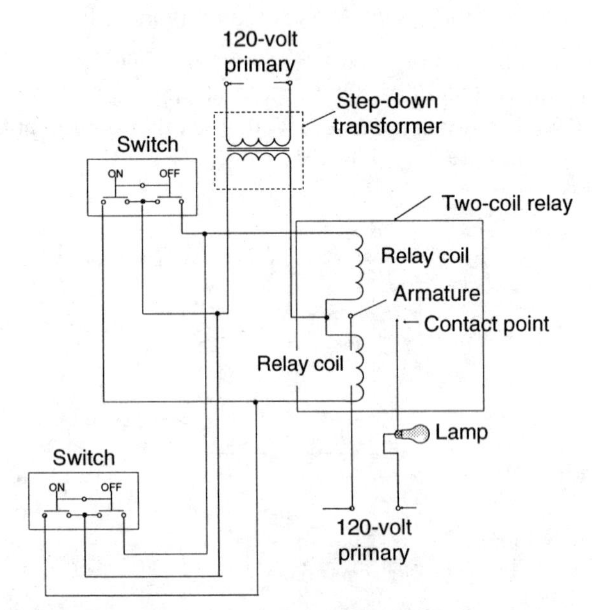

Figure 6-2: Basic remote-control switching circuit.

Labor units and material lists for most residential low-voltage installations are covered in this chapter.

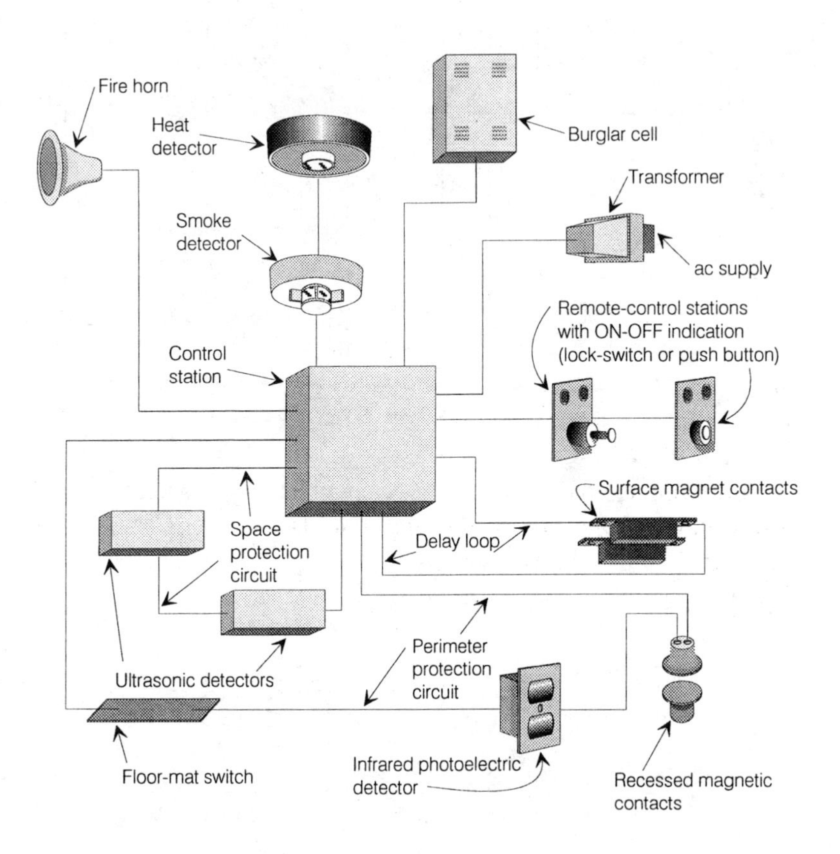

Figure 6-3: Components of a typical residential security/fire-alarm system.

Chime with One Pushbutton

Cost Item	Quantity	Price or Rate	Per	Installation Groups			
				1	2	3	4
MATERIAL							
Box and support	1		E				
Transformer	1		E				
Pushbutton	1		E				
#14-2 w/grd. Type NM cable	20 - 30 ft.		C Ft.				
#18 low-voltage cable	45 ft.		C Ft.				
Miscellaneous	Lot						
TOTAL MATERIAL COST							
TOTAL LABOR COST - GROUP 1	1.75 WH		Hr				
GROUP 2	2.50 WH		Hr				
GROUP 3	3.65 WH		Hr				
GROUP 4	4.80 WH		Hr				
DIRECT JOB EXPENSE							
TOTAL PRIME COST				$	$	$	$

Chime with Two Pushbuttons

Cost Item	Quantity	Price or Rate	Per	Installation Groups			
MATERIAL				1	2	3	4
Box and support	1		E				
Transformer	1		E				
Pushbuttons	2		E				
#14-2 w/grd. Type NM cable	20 - 30 ft.		C Ft.				
#18 low-voltage cable	80 ft.		C Ft.				
Miscellaneous	Lot						
TOTAL MATERIAL COST							
TOTAL LABOR COST - GROUP 1	2.35 WH		Hr				
GROUP 2	3.50 WH		Hr				
GROUP 3	4.65 WH		Hr				
GROUP 4	5.80 WH		Hr				
DIRECT JOB EXPENSE							
TOTAL PRIME COST				$	$	$	$

Relay and One Low-Voltage Switch

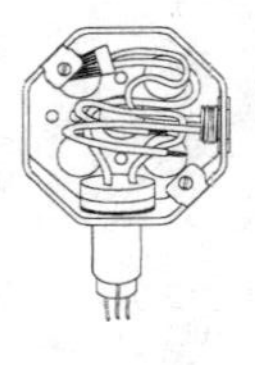

Cost Item	Quantity	Price or Rate	Per	Installation Groups			
MATERIAL				1	2	3	4
Plaster ring for switch box	1		E				
Low-voltage switch with cover	1		E				
#18-2 low-voltage cable	20 - 30 ft.		C Ft.				
#18-3 low-voltage cable	25 ft.		C Ft.				
Miscellaneous	Lot						
TOTAL MATERIAL COST							
TOTAL LABOR COST - GROUP 1	0.75 WH		Hr				
GROUP 2	1.15 WH		Hr				
GROUP 3	1.65 WH		Hr				
GROUP 4	1.80 WH		Hr				
DIRECT JOB EXPENSE							
TOTAL PRIME COST				$	$	$	$

Additional Low-Voltage Switch

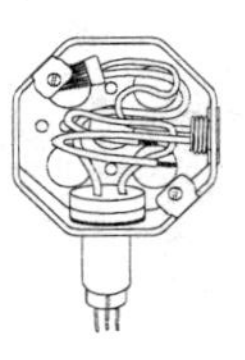

Cost Item	Quantity	Price or Rate	Per	Installation Groups			
MATERIAL				1	2	3	4
Plaster ring for switch box	1		E				
Low-voltage switch with cover	1		E				
#18-3 low-voltage cable	20 - 30 ft.		C Ft.				
Miscellaneous	Lot						
TOTAL MATERIAL COST							
TOTAL LABOR COST - GROUP 1	0.45 WH		Hr				
GROUP 2	0.70 WH		Hr				
GROUP 3	0.95 WH		Hr				
GROUP 4	1.30 WH		Hr				
DIRECT JOB EXPENSE							
TOTAL PRIME COST				$	$	$	$

Master Selector Switch

Cost Item	Quantity	Price or Rate	Per	Installation Groups			
				1	2	3	4
MATERIAL							
Box and support	1		E				
Master selector switch, complete	1		E				
#18-2 low-voltage cable	20 - 30 ft.		C Ft.				
#18-3 low-voltage cable	420 ft.		C Ft.				
Miscellaneous	Lot						
TOTAL MATERIAL COST							
TOTAL LABOR COST - GROUP 1	6.35 WH		Hr				
GROUP 2	9.50 WH		Hr				
GROUP 3	12.65 WH		Hr				
GROUP 4	14.80 WH		Hr				
DIRECT JOB EXPENSE							
TOTAL PRIME COST				$	$	$	$

Low-Voltage Supply Transformer

Cost Item	Quantity	Price or Rate	Per	Installation Groups 1	2	3	4
MATERIAL							
Box and support	1		E				
Transformer	1		E				
Miscellaneous	Lot						
TOTAL MATERIAL COST							
TOTAL LABOR COST - GROUP 1	0.35 WH		Hr				
GROUP 2	0.45 WH		Hr				
GROUP 3	0.65 WH		Hr				
GROUP 4	0.80 WH		Hr				
DIRECT JOB EXPENSE							
TOTAL PRIME COST				$	$	$	$

Key-Operated Delayed Entry Control

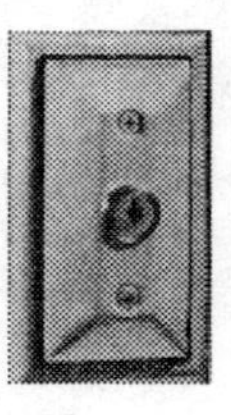

Cost Item	Quantity	Price or Rate	Per	Installation Groups			
MATERIAL				1	2	3	4
Box and support	1		E				
Low-voltage cable	20 - 30 ft.		M Ft.				
Key-operated delayed entry control	1		E				
Miscellaneous	Lot						
TOTAL MATERIAL COST							
TOTAL LABOR COST - GROUP 1	1.00 WH		Hr				
GROUP 2	1.50 WH		Hr				
GROUP 3	2.00 WH		Hr				
GROUP 4	2.80 WH		Hr				
DIRECT JOB EXPENSE							
TOTAL PRIME COST				$	$	$	$

Security-System Control Panel

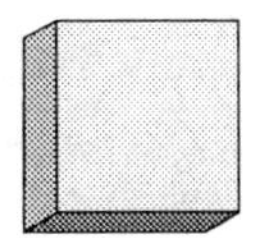

Cost Item	Quantity	Price or Rate	Per	Installation Groups			
MATERIAL				1	2	3	4
Control panel	1		E				
Miscellaneous	Lot						
TOTAL MATERIAL COST							
TOTAL LABOR COST - GROUP 1	1.75 WH		Hr				
GROUP 2	2.65 WH		Hr				
GROUP 3	3.65 WH		Hr				
GROUP 4	4.80 WH		Hr				
DIRECT JOB EXPENSE							
TOTAL PRIME COST				$	$	$	$

Stand-By Battery

Cost Item	Quantity	Price or Rate	Per	Installation Groups 1	2	3	4
MATERIAL							
Battery and housing	1		E				
Miscellaneous	Lot						
TOTAL MATERIAL COST							
TOTAL LABOR COST - GROUP 1	0.30 WH		Hr				
GROUP 2	0.50 WH		Hr				
GROUP 3	0.75 WH		Hr				
GROUP 4	0.85 WH		Hr				
DIRECT JOB EXPENSE							
TOTAL PRIME COST				$	$	$	$

Indoor Remote Station

Cost Item	Quantity	Price or Rate	Per	Installation Groups 1	2	3	4
MATERIAL							
Indoor remote station	1		E				
#18-2 low-voltage cable	60 ft.		C Ft.				
Miscellaneous	Lot						
TOTAL MATERIAL COST							
TOTAL LABOR COST - GROUP 1	2.50 WH		Hr				
GROUP 2	3.50 WH		Hr				
GROUP 3	4.65 WH		Hr				
GROUP 4	5.80 WH		Hr				
DIRECT JOB EXPENSE							
TOTAL PRIME COST				$	$	$	$

Prealarm Station

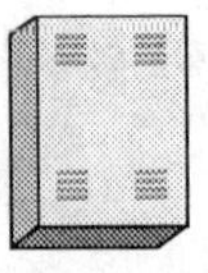

Cost Item	Quantity	Price or Rate	Per	Installation Groups 1	2	3	4
MATERIAL							
Prealarm station	1		E				
#18-2 low-voltage cable	20 - 30 ft.		C Ft.				
Miscellaneous	Lot						
TOTAL MATERIAL COST							
TOTAL LABOR COST - GROUP 1	0.50 WH		Hr				
GROUP 2	0.65 WH		Hr				
GROUP 3	0.80 WH		Hr				
GROUP 4	1.10 WH		Hr				
DIRECT JOB EXPENSE							
TOTAL PRIME COST				$	$	$	$

Outdoor Sounding Device

Cost Item	Quantity	Price or Rate	Per	Installation Groups			
				1	2	3	4
MATERIAL							
Outdoor sounding device	1		E				
#18-2 low-voltage cable	20 - 30 ft.		C Ft.				
Miscellaneous	Lot						
TOTAL MATERIAL COST							
TOTAL LABOR COST - GROUP 1	1.25 WH		Hr				
GROUP 2	1.50 WH		Hr				
GROUP 3	1.65 WH		Hr				
GROUP 4	1.80 WH		Hr				
DIRECT JOB EXPENSE							
TOTAL PRIME COST				$	$	$	$

Indoor Sounding Device

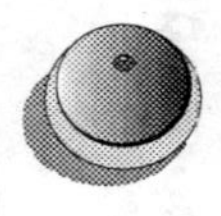

Cost Item	Quantity	Price or Rate	Per	Installation Groups			
				1	2	3	4
MATERIAL							
Indoor sounding device	1		E				
#18-2 low-voltage cable	25 ft.		C Ft.				
Miscellaneous	Lot						
TOTAL MATERIAL COST							
TOTAL LABOR COST - GROUP 1	0.70 WH		Hr				
GROUP 2	1.00 WH		Hr				
GROUP 3	1.30 WH		Hr				
GROUP 4	1.60 WH		Hr				
DIRECT JOB EXPENSE							
TOTAL PRIME COST				$	$	$	$

Siren Driver

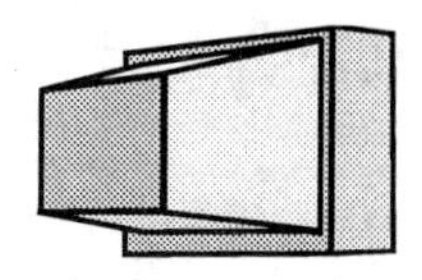

Cost Item	Quantity	Price or Rate	Per	Installation Groups			
				1	2	3	4
MATERIAL							
Siren driver	1		E				
#14-2 w/grd. Type NM cable	20 - 30 ft.		C Ft.				
#18-2 low-voltage cable	45 ft.		C Ft.				
Miscellaneous	Lot						
TOTAL MATERIAL COST							
TOTAL LABOR COST - GROUP 1	1.00 WH		Hr				
GROUP 2	1.50 WH		Hr				
GROUP 3	2.00 WH		Hr				
GROUP 4	2.80 WH		Hr				
DIRECT JOB EXPENSE							
TOTAL PRIME COST				$	$	$	$

Relay

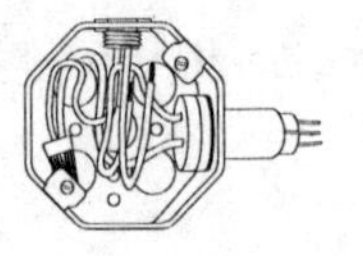

Cost Item	Quantity	Price or Rate	Per	Installation Groups 1	2	3	4
MATERIAL							
Box and support	1		E				
Low-voltage relay	1						
Miscellaneous	Lot						
TOTAL MATERIAL COST							
TOTAL LABOR COST - GROUP 1	1.20 WH		Hr				
GROUP 2	1.50 WH		Hr				
GROUP 3	2.65 WH		Hr				
GROUP 4	2.80 WH		Hr				
DIRECT JOB EXPENSE							
TOTAL PRIME COST				$	$	$	$

Photoelectric System

Cost Item	Quantity	Price or Rate	Per	Installation Groups			
MATERIAL				1	2	3	4
Photoelectric system	1		E				
#18-2 low-voltage cable	20 - 30 ft.		C Ft.				
Miscellaneous	Lot						
TOTAL MATERIAL COST							
TOTAL LABOR COST - GROUP 1	1.20 WH		Hr				
GROUP 2	1.90 WH		Hr				
GROUP 3	2.65 WH		Hr				
GROUP 4	3.80 WH		Hr				
DIRECT JOB EXPENSE							
TOTAL PRIME COST				$	$	$	$

Smoke Detector

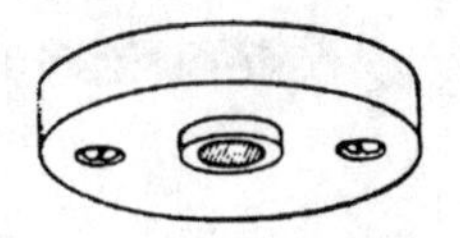

Cost Item	Quantity	Price or Rate	Per	Installation Groups 1	2	3	4
MATERIAL							
Box and support	1		E				
Smoke detector	1		C Ft.				
Miscellaneous	Lot						
TOTAL MATERIAL COST							
TOTAL LABOR COST - GROUP 1	0.80 WH		Hr				
GROUP 2	1.10 WH		Hr				
GROUP 3	1.30 WH		Hr				
GROUP 4	1.80 WH		Hr				
DIRECT JOB EXPENSE							
TOTAL PRIME COST				$	$	$	$

Magnetic Contacts

Cost Item	Quantity	Price or Rate	Per	Installation Groups			
				1	2	3	4
MATERIAL							
Magnetic contacts	2		E				
#18-2 low-voltage cable	25 ft.		C Ft.				
Miscellaneous	Lot						
TOTAL MATERIAL COST							
TOTAL LABOR COST - GROUP 1	0.40 WH		Hr				
GROUP 2	0.55 WH		Hr				
GROUP 3	0.75 WH		Hr				
GROUP 4	0.80 WH		Hr				
DIRECT JOB EXPENSE							
TOTAL PRIME COST				$	$	$	$

Glassbreak Detector

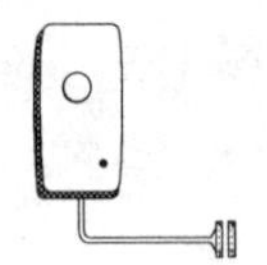

Cost Item	Quantity	Price or Rate	Per	Installation Groups			
				1	2	3	4
MATERIAL							
Glassbreak detector	1		E				
#18-2 low-voltage cable	30 ft.		C Ft.				
Miscellaneous	Lot						
TOTAL MATERIAL COST							
TOTAL LABOR COST - GROUP 1	0.70 WH		Hr				
GROUP 2	1.00 WH		Hr				
GROUP 3	1.30 WH		Hr				
GROUP 4	1.60 WH		Hr				
DIRECT JOB EXPENSE							
TOTAL PRIME COST				$	$	$	$

Emergency Switch

Cost Item	Quantity	Price or Rate	Per	Installation Groups 1	2	3	4
MATERIAL							
Box and support	1		E				
Emergency switch	1		E				
#18-2 low-voltage cable	20 - 30 ft.		C Ft.				
Miscellaneous	Lot						
TOTAL MATERIAL COST							
TOTAL LABOR COST - GROUP 1	1.00 WH		Hr				
GROUP 2	1.50 WH		Hr				
GROUP 3	2.00 WH		Hr				
GROUP 4	3.00 WH		Hr				
DIRECT JOB EXPENSE							
TOTAL PRIME COST				$	$	$	$

Low-Voltage HVAC Thermostat

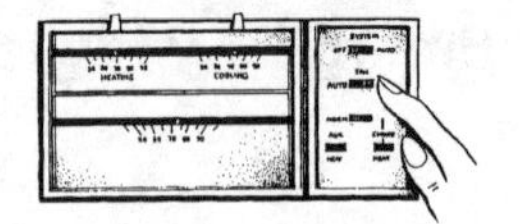

Cost Item	Quantity	Price or Rate	Per	Installation Groups 1	2	3	4
MATERIAL							
Low-voltage thermostat	1		E				
#18-3 low-voltage cable	50 ft.		C Ft.				
Miscellaneous	Lot						
TOTAL MATERIAL COST							
TOTAL LABOR COST - GROUP 1	1.35 WH		Hr				
GROUP 2	2.50 WH		Hr				
GROUP 3	3.65 WH		Hr				
GROUP 4	4.80 WH		Hr				
DIRECT JOB EXPENSE							
TOTAL PRIME COST				$	$	$	$

Selling-Price Tables

The selling-price tables to follow are designed for quick unit pricing of the various low-voltage outlets found in this chapter. Before arriving at a selling price, the preceding tables must be completed (filled in with appropriate prices). Then the prime cost determined for each wiring situation in the preceding tables should be entered in the corresponding selling-price tables to follow. Once the prime cost has been entered in each selling-price table, overhead and profit factors are calculated to arrive at a total unit selling price as described in detail in Chapter 2.

Chime with One Pushbutton		**GROUPS**			
		1	**2**	**3**	**4**
Total Prime Cost		**$**	**$**	**$**	**$**
Overhead	_% of prime cost				
Total Cost		**$**	**$**	**$**	**$**
Profit	_% of total cost				
Selling Price		**$**	**$**	**$**	**$**

Chime with Two Pushbuttons		**GROUPS**			
		1	**2**	**3**	**4**
Total Prime Cost		**$**	**$**	**$**	**$**
Overhead	_% of prime cost				
Total Cost		**$**	**$**	**$**	**$**
Profit	_% of total cost				
Selling Price		**$**	**$**	**$**	**$**

Relay and One Low-Voltage Switch		**GROUPS**			
		1	**2**	**3**	**4**
Total Prime Cost		**$**	**$**	**$**	**$**
Overhead	_% of prime cost				
Total Cost		**$**	**$**	**$**	**$**
Profit	_% of total cost				
Selling Price		**$**	**$**	**$**	**$**

Additional Low-Voltage Switch		GROUPS			
		1	2	3	4
Total Prime Cost		$	$	$	$
Overhead	_% of prime cost				
Total Cost		$	$	$	$
Profit	_% of total cost				
Selling Price		$	$	$	$

Master Selector Switch		GROUPS			
		1	2	3	4
Total Prime Cost		$	$	$	$
Overhead	_% of prime cost				
Total Cost		$	$	$	$
Profit	_% of total cost				
Selling Price		$	$	$	$

Low-Voltage Supply Transformer		GROUPS			
		1	2	3	4
Total Prime Cost		$	$	$	$
Overhead	_% of prime cost				
Total Cost		$	$	$	$
Profit	_% of total cost				
Selling Price		$	$	$	$

Key-Operated Delayed Entry Control		**GROUPS**			
		1	**2**	**3**	**4**
Total Prime Cost		**$**	**$**	**$**	**$**
Overhead	_% of prime cost				
Total Cost		**$**	**$**	**$**	**$**
Profit	_% of total cost				
Selling Price		**$**	**$**	**$**	**$**

Security-System Control Panel		**GROUPS**			
		1	**2**	**3**	**4**
Total Prime Cost		**$**	**$**	**$**	**$**
Overhead	_% of prime cost				
Total Cost		**$**	**$**	**$**	**$**
Profit	_% of total cost				
Selling Price		**$**	**$**	**$**	**$**

Stand-By Battery		**GROUPS**			
		1	**2**	**3**	**4**
Total Prime Cost		**$**	**$**	**$**	**$**
Overhead	_% of prime cost				
Total Cost		**$**	**$**	**$**	**$**
Profit	_% of total cost				
Selling Price		**$**	**$**	**$**	**$**

Indoor Remote Station		GROUPS			
		1	2	3	4
Total Prime Cost		$	$	$	$
Overhead	_% of prime cost				
Total Cost		$	$	$	$
Profit	_% of total cost				
Selling Price		$	$	$	$

Prealarm Station		GROUPS			
		1	2	3	4
Total Prime Cost		$	$	$	$
Overhead	_% of prime cost				
Total Cost		$	$	$	$
Profit	_% of total cost				
Selling Price		$	$	$	$

Outdoor Sounding Device		GROUPS			
		1	2	3	4
Total Prime Cost		$	$	$	$
Overhead	_% of prime cost				
Total Cost		$	$	$	$
Profit	_% of total cost				
Selling Price		$	$	$	$

Indoor Sounding Device		GROUPS			
		1	2	3	4
Total Prime Cost		$	$	$	$
Overhead	_% of prime cost				
Total Cost		$	$	$	$
Profit	_% of total cost				
Selling Price		$	$	$	$

Siren Driver		GROUPS			
		1	2	3	4
Total Prime Cost		$	$	$	$
Overhead	_% of prime cost				
Total Cost		$	$	$	$
Profit	_% of total cost				
Selling Price		$	$	$	$

Relay		GROUPS			
		1	2	3	4
Total Prime Cost		$	$	$	$
Overhead	_% of prime cost				
Total Cost		$	$	$	$
Profit	_% of total cost				
Selling Price		$	$	$	$

Photoelectric System		GROUPS			
		1	2	3	4
Total Prime Cost		$	$	$	$
Overhead	_% of prime cost				
Total Cost		$	$	$	$
Profit	_% of total cost				
Selling Price		$	$	$	$

Smoke Detector		GROUPS			
		1	2	3	4
Total Prime Cost		$	$	$	$
Overhead	_% of prime cost				
Total Cost		$	$	$	$
Profit	_% of total cost				
Selling Price		$	$	$	$

Magnetic Contacts		GROUPS			
		1	2	3	4
Total Prime Cost		$	$	$	$
Overhead	_% of prime cost				
Total Cost		$	$	$	$
Profit	_% of total cost				
Selling Price		$	$	$	$

Glassbreak Detector		GROUPS			
		1	2	3	4
Total Prime Cost		$	$	$	$
Overhead	_% of prime cost				
Total Cost		$	$	$	$
Profit	_% of total cost				
Selling Price		$	$	$	$

Emergency Switch		GROUPS			
		1	2	3	4
Total Prime Cost		$	$	$	$
Overhead	_% of prime cost				
Total Cost		$	$	$	$
Profit	_% of total cost				
Selling Price		$	$	$	$

Low-Voltage HVAC Thermostat		GROUPS			
		1	2	3	4
Total Prime Cost		$	$	$	$
Overhead	_% of prime cost				
Total Cost		$	$	$	$
Profit	_% of total cost				
Selling Price		$	$	$	$

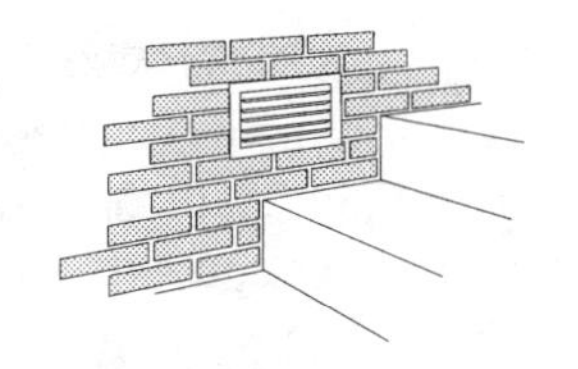

Chapter 7

Outdoor Wiring

Items covered in this chapter include conventional outdoor lighting, switches, and receptacles used in residential applications. These items may be further subdivided into the following:

- Post lights (Figure 7-1 on next page)
- Mushroom lights
- Floodlighting
- Step lights
- Underwater pool or fountain lights
- Weatherproof receptacles
- Weatherproof switches
- Mercury-vapor lighting fixtures

Although overhead wiring is still used to some extent, underground wiring is the most popular wiring method as it requires very little (if any) maintenance, is not harmed by winter ice and snow storms, and is usually the least expensive to install. Type U.F. (underground feeder) cable is the least expensive and most popular wiring method used for residential underground wiring, but PVC conduit is also frequently used for underground wiring in residential applications. Where U.F. cable extends above grade, some means of mechanical protection is necessary; that is, the cable must be enclosed in a raceway or else protected in some other way.

All outside receptacles within 8 feet of grade level must be protected by a ground-fault circuit-interrupter to comply with *NEC* requirements. All device covers must be weatherproof, and all outlet boxes, conductors, and other circuit components must be designed for outdoor use.

Underwater fountain and pool lighting require special wiring methods and must comply with all provisions stated in *NEC* Article 680.

No estimating charts in this chapter include the cost of lighting fixtures.

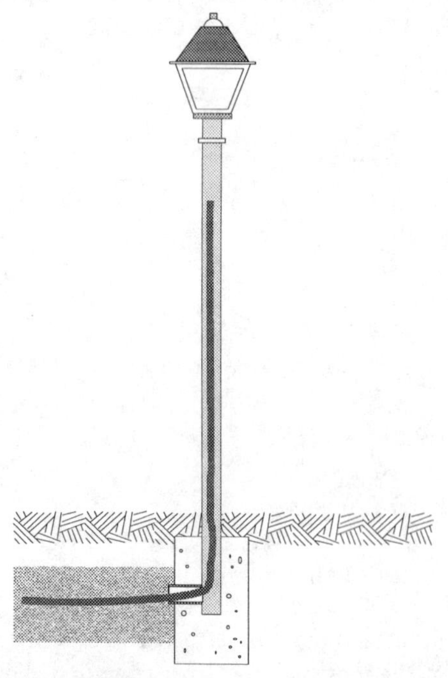

Figure 7-1: Lighting standards or post lights are common at sidewalk entrances to many residences.

Outdoor Step Light

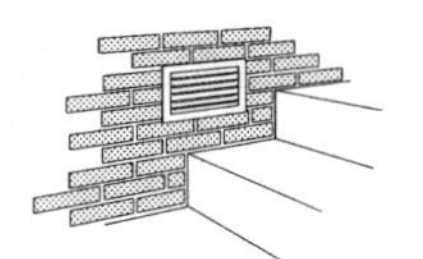

Cost Item	Quantity	Price or Rate	Per	Installation Groups			
				1	2	3	4
MATERIAL							
Rigid PVC conduit	12		C Ft.				
PVC entrance elbow or LB conduit body	1		E				
#12-2 w/ground Type U.F. cable	45 ft.		C Ft.				
Miscellaneous	Lot						
TOTAL MATERIAL COST							
TOTAL LABOR COST - GROUP 1	7.00 WH		Hr				
GROUP 2	10.50 WH		Hr				
GROUP 3	14.65 WH		Hr				
GROUP 4	18.80 WH		Hr				
DIRECT JOB EXPENSE							
TOTAL PRIME COST				$	$	$	$

Outdoor Floodlight

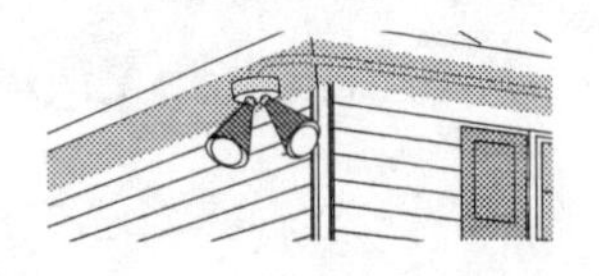

Cost Item	Quantity	Price or Rate	Per	Installation Groups 1	2	3	4
MATERIAL							
Weatherproof outlet box	1		E				
#12-2 w/ground Type U.F. cable	50 ft.		C Ft.				
Miscellaneous	Lot						
TOTAL MATERIAL COST							
TOTAL LABOR COST - GROUP 1	3.35 WH		Hr				
GROUP 2	4.50 WH		Hr				
GROUP 3	5.65 WH		Hr				
GROUP 4	6.80 WH		Hr				
DIRECT JOB EXPENSE							
TOTAL PRIME COST				$	$	$	$

Outdoor Post Lighting Outlet

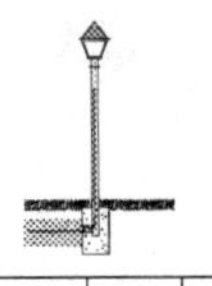

Cost Item	Quantity	Price or Rate	Per	Installation Groups			
				1	2	3	4
MATERIAL							
Weatherproof junction box	1		E				
PVC rigid conduit	20 ft.						
#12-2 w/ground Type U.F. cable	50 ft.		C Ft.				
Miscellaneous	Lot						
TOTAL MATERIAL COST							
TOTAL LABOR COST - GROUP 1	5.35 WH		Hr				
GROUP 2	8.50 WH		Hr				
GROUP 3	11.65 WH		Hr				
GROUP 4	14.80 WH		Hr				
DIRECT JOB EXPENSE							
TOTAL PRIME COST				$	$	$	$

Outdoor Mushroom Lighting Outlet

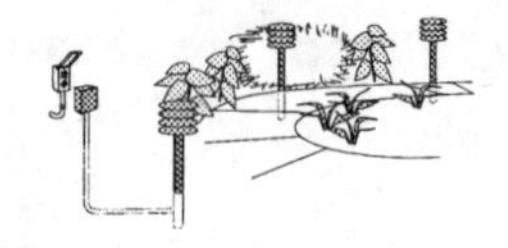

Cost Item	Quantity	Price or Rate	Per	Installation Groups 1	2	3	4
MATERIAL							
Weatherproof outlet box (if needed)	1		E				
PVC rigid conduit	10 ft.						
#12-2 w/ground Type U.F. cable	40 ft.		C Ft.				
Miscellaneous	Lot						
TOTAL MATERIAL COST							
TOTAL LABOR COST - GROUP 1	5.00 WH		Hr				
GROUP 2	8.00 WH		Hr				
GROUP 3	11.00 WH		Hr				
GROUP 4	14.00 WH		Hr				
DIRECT JOB EXPENSE							
TOTAL PRIME COST				$	$	$	$

Underwater Pool or Fountain Lighting Outlet

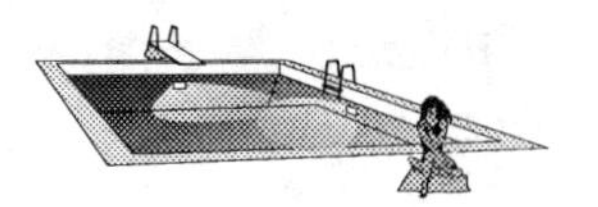

Cost Item	Quantity	Price or Rate	Per	Installation Groups			
				1	2	3	4
MATERIAL							
#12/3 MI cable	40 ft.		C Ft.				
MI cable connectors	2		E				
Miscellaneous	Lot						
TOTAL MATERIAL COST							
TOTAL LABOR COST - GROUP 1	7.35 WH		Hr				
GROUP 2	9.50 WH		Hr				
GROUP 3	11.65 WH		Hr				
GROUP 4	17.80 WH		Hr				
DIRECT JOB EXPENSE							
TOTAL PRIME COST				$	$	$	$

Outdoor Weatherproof Receptacle

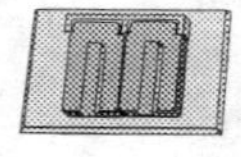

Cost Item	Quantity	Price or Rate	Per	Installation Groups			
				1	2	3	4
MATERIAL							
Weatherproof outlet box	1		E				
PVC rigid conduit	10 ft.		C Ft.				
Entrance elbow or LP conduit body	1		E				
Duplex receptacle with raintight cover	1		E				
#12-2 w/ground Type U.F. cable	40 ft.		C Ft.				
Miscellaneous	Lot						
TOTAL MATERIAL COST							
TOTAL LABOR COST - GROUP 1	6.35 WH		Hr				
GROUP 2	10.50 WH		Hr				
GROUP 3	14.65 WH		Hr				
GROUP 4	17.80 WH		Hr				
DIRECT JOB EXPENSE							
TOTAL PRIME COST				$	$	$	$

Outdoor Weatherproof Switch

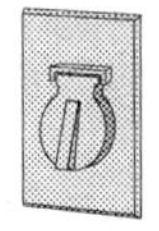

Cost Item	Quantity	Price or Rate	Per	Installation Groups 1	2	3	4
MATERIAL							
Weatherproof outlet box	1		E				
Entrance elbow or LP conduit body	1		E				
Switch with weatherproof cover	1		E				
#12-2 w/ground Type U.F. cable	50 ft.		M Ft.				
Miscellaneous	Lot						
TOTAL MATERIAL COST							
TOTAL LABOR COST - GROUP 1	6.35 WH		Hr				
GROUP 2	9.50 WH		Hr				
GROUP 3	13.65 WH		Hr				
GROUP 4	16.80 WH		Hr				
DIRECT JOB EXPENSE							

Outdoor Dusk 'Til Dawn Mercury-Vapor Lighting Outlet

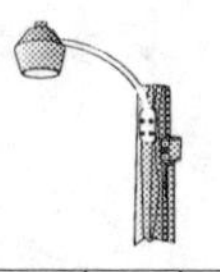

Cost Item	Quantity	Price or Rate	Per	Installation Groups			
				1	2	3	4
MATERIAL							
Weatherproof junction box w/cover	1		E				
PVC rigid conduit	20 ft.						
#12-2 w/ground Type U.F. cable	100 ft.		C Ft.				
Miscellaneous	Lot						
TOTAL MATERIAL COST							
TOTAL LABOR COST - GROUP 1	8.00 WH		Hr				
GROUP 2	12.00 WH		Hr				
GROUP 3	16.00 WH		Hr				
GROUP 4	20.00 WH		Hr				
DIRECT JOB EXPENSE							
TOTAL PRIME COST				$	$	$	$

Selling-Price Tables

The selling-price tables to follow are designed for quick unit pricing of the various outdoor outlets found in this chapter. Before arriving at a selling price, the preceding tables must be completed (filled in with appropriate prices). Then the prime cost determined for each wiring situation in the preceding tables should be entered in the corresponding selling-price tables to follow. Once the prime cost has been entered in each selling-price table, overhead and profit factors are calculated to arrive at a total unit selling price as described in detail in Chapter 2.

Outdoor Step Light		**GROUPS**			
		1	**2**	**3**	**4**
Total Prime Cost		$	$	$	$
Overhead	_% of prime cost				
Total Cost		$	$	$	$
Profit	_% of total cost				
Selling Price		$	$	$	$

Outdoor Floodlight		**GROUPS**			
		1	**2**	**3**	**4**
Total Prime Cost		$	$	$	$
Overhead	_% of prime cost				
Total Cost		$	$	$	$
Profit	_% of total cost				
Selling Price		$	$	$	$

Outdoor Post Lighting Outlet		**GROUPS**			
		1	**2**	**3**	**4**
Total Prime Cost		$	$	$	$
Overhead	_% of prime cost				
Total Cost		$	$	$	$
Profit	_% of total cost				
Selling Price		$	$	$	$

Outdoor Mushroom Lighting Outlet		GROUPS			
		1	2	3	4
Total Prime Cost		$	$	$	$
Overhead	_% of prime cost				
Total Cost		$	$	$	$
Profit	_% of total cost				
Selling Price		$	$	$	$

Underwater Pool or Fountain Lighting Outlet		GROUPS			
		1	2	3	4
Total Prime Cost		$	$	$	$
Overhead	_% of prime cost				
Total Cost		$	$	$	$
Profit	_% of total cost				
Selling Price		$	$	$	$

Outdoor Weatherproof Receptacle		GROUPS			
		1	2	3	4
Total Prime Cost		$	$	$	$
Overhead	_% of prime cost				
Total Cost		$	$	$	$
Profit	_% of total cost				
Selling Price		$	$	$	$

Outdoor Weatherproof Switch		GROUPS			
		1	2	3	4
Total Prime Cost		$	$	$	$
Overhead	_% of prime cost				
Total Cost		$	$	$	$
Profit	_% of total cost				
Selling Price		$	$	$	$

Outdoor Dusk 'Til Dawn Mercury-Vapor Lighting Outlet		GROUPS			
		1	2	3	4
Total Prime Cost		$	$	$	$
Overhead	_% of prime cost				
Total Cost		$	$	$	$
Profit	_% of total cost				
Selling Price		$	$	$	$

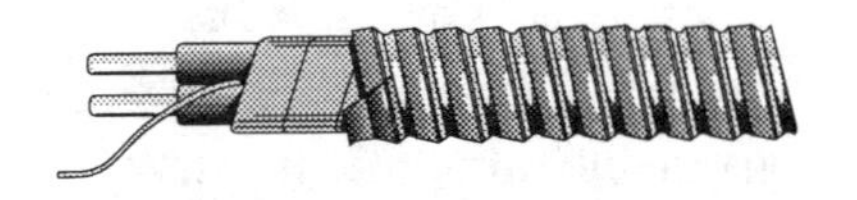

Chapter 8

Alternate Residential Wiring Methods

Most residential electrical systems seldom use only one wiring method throughout. For example, interior lighting, receptacle, and switch outlets may use Type NM cable. Rigid steel conduit may be used for the service mast and also for wiring in a shop area in the basement or garage. PVC rigid conduit and Type UF cable may be used for exterior wiring, while EMT may be used for part of the basement wiring of wall-mounted switches and receptacles. The combinations are almost endless.

In most cases, the number of outlets utilizing these alternate wiring methods is minute when compared to the overall installation. Therefore, these outlets do not lend themselves to satisfactory unit pricing. There is, however, one exception: Armored or Type AC cable, commonly called "BX" on the job, is the preferred wiring method for residential occupancies that are more than three stories high; the *NEC* does not allow Type NM cable in such installations. Furthermore, to obtain greater mechanical protection, many architects and consulting engineers specify Type AC cable as opposed to Type NM cable.

Consequently, complete pricing charts have been included in this chapter for Type AC cable.

Other alternate wiring methods should be priced according to the given situation; that is, make a material takeoff, calculate the total worker-hours, and price all of these items to obtain a prime cost of the given installation. Then add overhead and profit to obtain a selling price as usual.

Type AC (Armored Cable)

Type AC (armored) cable is manufactured in two-, three-, and four-wire assemblies, with varying sizes of conductors, and is used in locations similar to those where Type NM (nonmetallic) cable is allowed. The metallic spiral covering on BX cable offers a greater degree of mechanical protection than with NM cable, and the metal jacket also provides a continuous grounding bond without the need for additional grounding conductors.

BX cable may be used for under-plaster extensions, as provided in the *NEC*, and embedded in plaster finish, brick, or other masonry, except in damp or wet locations. It may also be run or "fished" in the air voids of masonry block or tile walls, except where such walls are exposed or subject to excessive moisture or dampness or are below grade. Furthermore, Type AC cable may not be used in theaters, places of assembly, in motion picture studios, in any hazardous location, in areas where the cable will be exposed to corrosive fumes or vapors, on cranes or hoists, and in commercial garages.

While most electrical contractors specializing in residential work rely mostly on Type NM cable for the major wiring method, BX cable must be used in apartment complexes where the building exceeds more than three floors. Furthermore, this cable is very handy when wiring existing structures where much "fishing" must be done. BX cable may be fished through partitions much easier than Type NM cable, and many electical contractors find that the added cost in materials is offset by the quicker installation time required to fish BX cable over Type NM cable. This cable also provides added protection in workshops and similar locations. In some cases, the owner, architect, or engineer may require Type AC cable to be used, with no substitution.

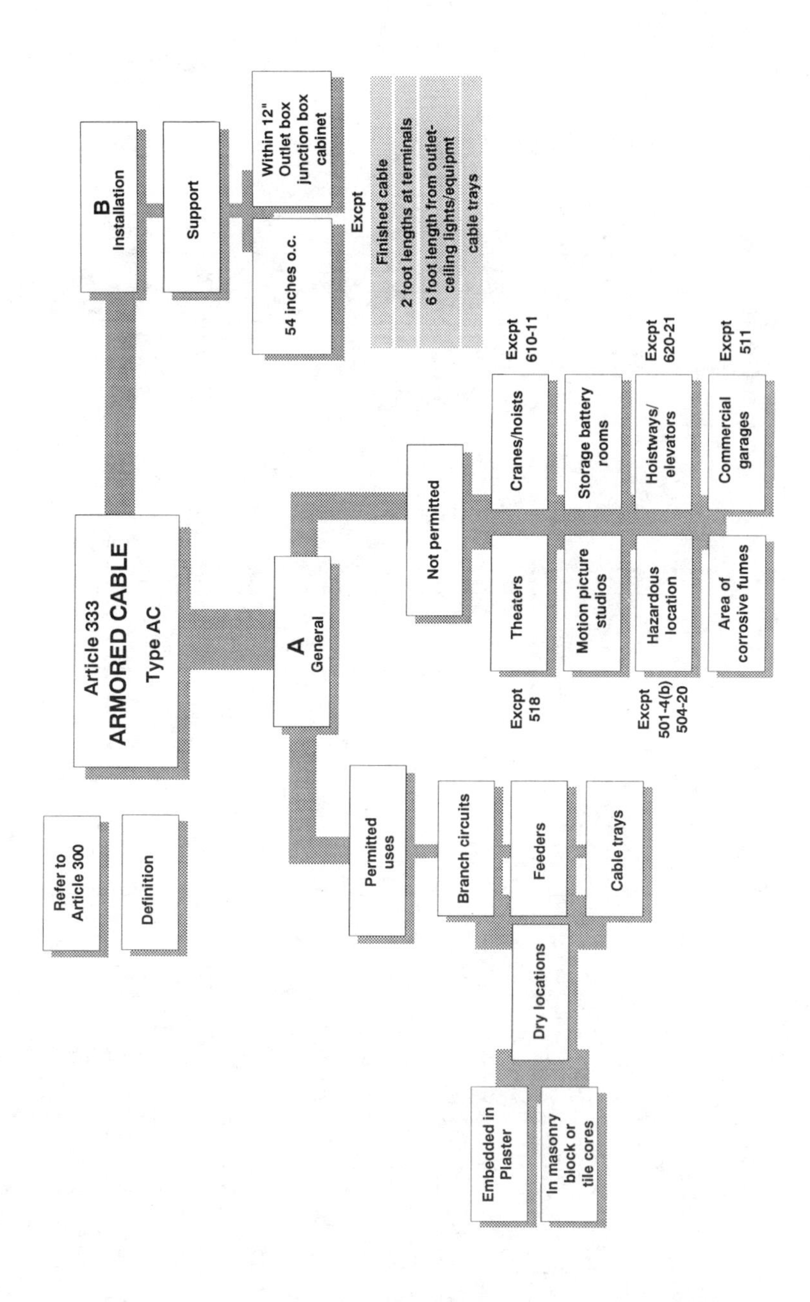

Figure 8-1: Summary of *NEC* regulations governing Type AC cable.

Surface-Mounted Lighting Outlet

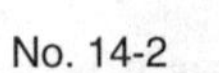

No. 14-2

Cost Item	Quantity	Price or Rate	Per	Installation Groups 1	2	3	4
MATERIAL							
Box and support	1		E				
#14-2 Type AC Cable	20 - 30 ft.		C Ft.				
Miscellaneous	Lot						
TOTAL MATERIAL COST							
TOTAL LABOR COST - GROUP 1	0.35 WH		Hr				
GROUP 2	0.45 WH		Hr				
GROUP 3	0.75 WH		Hr				
GROUP 4	1.00 WH		Hr				
DIRECT JOB EXPENSE							
TOTAL PRIME COST				$	$	$	$

Surface-Mounted Lighting Outlet

Cost Item	Quantity	Price or Rate	Per	Installation Groups 1	2	3	4
MATERIAL							
Box and support	1		E				
#12-2 Type AC Cable	20 - 30 ft.		C Ft.				
Miscellaneous	Lot						
TOTAL MATERIAL COST							
TOTAL LABOR COST - GROUP 1	0.45 WH		Hr				
GROUP 2	0.65 WH		Hr				
GROUP 3	0.85 WH		Hr				
GROUP 4	1.10 WH		Hr				
DIRECT JOB EXPENSE							
TOTAL PRIME COST				$	$	$	$

Recessed Lighting Outlet

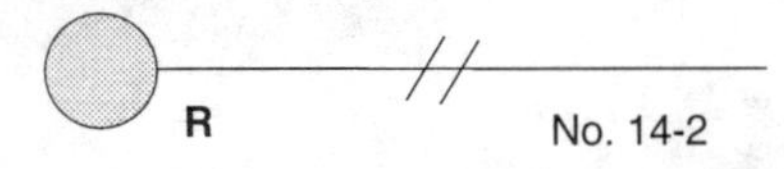

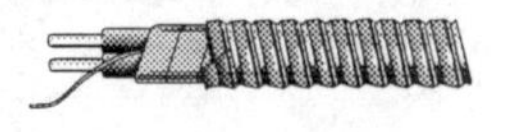

Cost Item	Quantity	Price or Rate	Per	Installation Groups 1	2	3	4
MATERIAL							
Cable connector	1		E				
#14-2 Type AC Cable	20 - 30 ft.		C Ft.				
Miscellaneous	Lot						
TOTAL MATERIAL COST							
TOTAL LABOR COST - GROUP 1	0.55 WH		Hr				
GROUP 2	0.85 WH		Hr				
GROUP 3	1.10 WH		Hr				
GROUP 4	1.45 WH		Hr				
DIRECT JOB EXPENSE							
TOTAL PRIME COST				$	$	$	$

Recessed Lighting Outlet

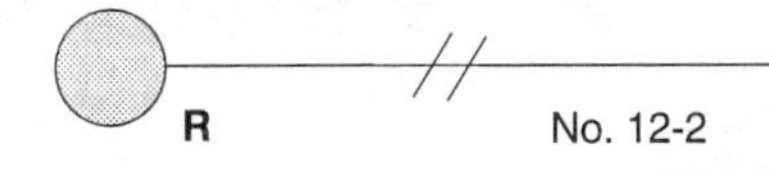

Cost Item	Quantity	Price or Rate	Per	Installation Groups 1	2	3	4
MATERIAL							
Cable connector	1		E				
#12-2 Type AC Cable	20 - 30 ft.		C Ft.				
Miscellaneous	Lot						
TOTAL MATERIAL COST							
TOTAL LABOR COST - GROUP 1	0.65 WH		Hr				
GROUP 2	0.95 WH		Hr				
GROUP 3	1.25 WH		Hr				
GROUP 4	1.55 WH		Hr				
DIRECT JOB EXPENSE							
TOTAL PRIME COST				$	$	$	$

Wall-Mounted Lighting Outlet

No. 14-2

Cost Item	Quantity	Price or Rate	Per	Installation Groups			
				1	2	3	4
MATERIAL							
Box and support	1		E				
#14-2 Type AC Cable	20 - 30 ft.		C Ft.				
Miscellaneous	Lot						
TOTAL MATERIAL COST							
TOTAL LABOR COST - GROUP 1	0.35 WH		Hr				
GROUP 2	0.55 WH		Hr				
GROUP 3	0.75 WH		Hr				
GROUP 4	0.95 WH		Hr				
DIRECT JOB EXPENSE							
TOTAL PRIME COST				$	$	$	$

Wall-Mounted Lighting Outlet

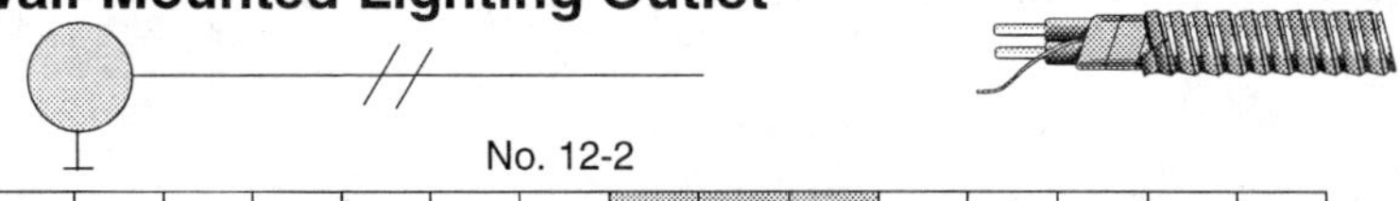

Cost Item	Quantity	Price or Rate	Per	Installation Groups			
				1	2	3	4
MATERIAL							
Box and support	1		E				
#12-2 Type AC Cable	20 - 30 ft.		C Ft.				
Miscellaneous	Lot						
TOTAL MATERIAL COST							
TOTAL LABOR COST - GROUP 1	0.45 WH		Hr				
GROUP 2	0.65 WH		Hr				
GROUP 3	0.85 WH		Hr				
GROUP 4	1.05 WH		Hr				
DIRECT JOB EXPENSE							
TOTAL PRIME COST				$	$	$	$

Duplex Receptacle – 2-Wire Circuit

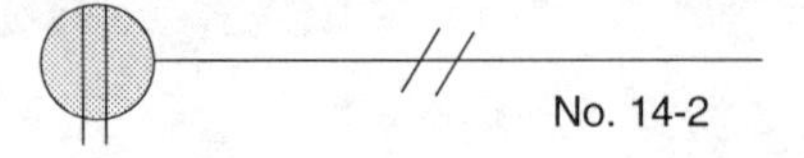

Cost Item	Quantity	Price or Rate	Per	Installation Groups 1	2	3	4
MATERIAL							
Box and support	1		E				
#14-2 Type AC Cable	20 - 30 ft.		C Ft.				
Duplex receptacle and plate	1		E				
Miscellaneous	Lot						
TOTAL MATERIAL COST							
TOTAL LABOR COST - GROUP 1	0.55 WH		Hr				
GROUP 2	0.85 WH		Hr				
GROUP 3	1.15 WH		Hr				
GROUP 4	1.45 WH		Hr				
DIRECT JOB EXPENSE							
TOTAL PRIME COST				$	$	$	$

Duplex Receptacle – 2-Wire Circuit

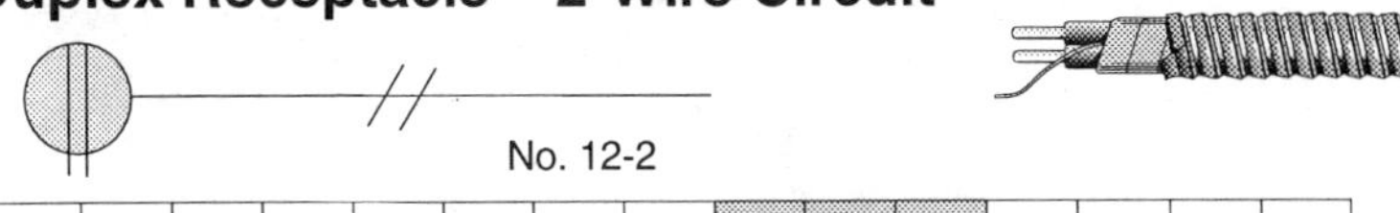

Cost Item	Quantity	Price or Rate	Per	Installation Groups			
MATERIAL				1	2	3	4
Box and support	1		E				
#12-2 Type AC Cable	20 - 30 ft.		C Ft.				
Duplex receptacle and plate	1		E				
Miscellaneous	Lot						
TOTAL MATERIAL COST							
TOTAL LABOR COST - GROUP 1	0.65 WH		Hr				
GROUP 2	0.95 WH		Hr				
GROUP 3	1.25 WH		Hr				
GROUP 4	1.55 WH		Hr				
DIRECT JOB EXPENSE							
TOTAL PRIME COST				$	$	$	$

Receptacle – 3-Wire Circuit

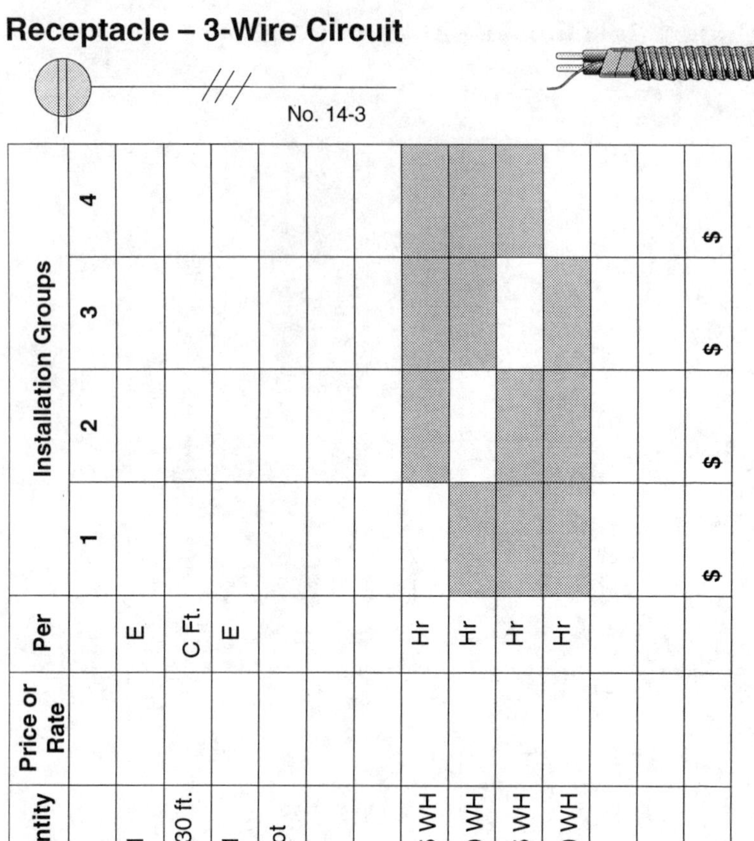

Cost Item	Quantity	Price or Rate	Per	Installation Groups 1	2	3	4
MATERIAL							
Box and support	1		E				
#14-3 Type AC Cable	20 - 30 ft.		C Ft.				
Receptacle and plate	1		E				
Miscellaneous	Lot						
TOTAL MATERIAL COST							
TOTAL LABOR COST - GROUP 1	0.65 WH		Hr				
GROUP 2	1.00 WH		Hr				
GROUP 3	1.35 WH		Hr				
GROUP 4	1.70 WH		Hr				
DIRECT JOB EXPENSE							
TOTAL PRIME COST				$	$	$	$

Receptacle – 3-Wire Circuit

Cost Item	Quantity	Price or Rate	Per	Installation Groups			
				1	2	3	4
MATERIAL							
Box and support	1		E				
#12-3 Type AC Cable	20 - 30 ft.		C Ft.				
Receptacle and Plate	1		E				
Miscellaneous	Lot						
TOTAL MATERIAL COST							
TOTAL LABOR COST - GROUP 1	0.75 WH		Hr				
GROUP 2	1.10 WH		Hr				
GROUP 3	1.45 WH		Hr				
GROUP 4	1.80 WH		Hr				
DIRECT JOB EXPENSE							
TOTAL PRIME COST				$	$	$	$

Duplex Receptacle – Split-Wired

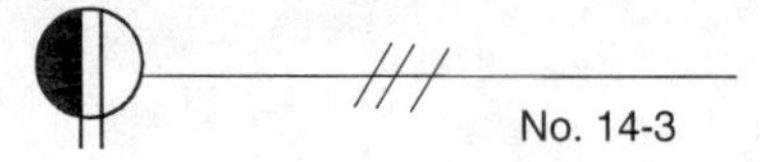

No. 14-3

Cost Item	Quantity	Price or Rate	Per	Installation Groups 1	2	3	4
MATERIAL							
Box and support	1		E				
#14-3 Type AC Cable	20 - 30 ft.	.	C Ft.				
Two-circuit receptacle and plate	1		E				
Miscellaneous	Lot						
TOTAL MATERIAL COST							
TOTAL LABOR COST - GROUP 1	0.65 WH		Hr				
GROUP 2	1.00 WH		Hr				
GROUP 3	1.35 WH		Hr				
GROUP 4	1.70 WH		Hr				
DIRECT JOB EXPENSE							
TOTAL PRIME COST				$	$	$	$

Duplex Receptacle – Split-Wired

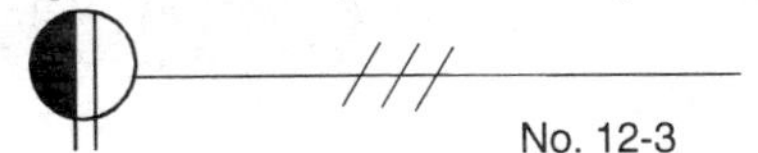

No. 12-3

Cost Item	Quantity	Price or Rate	Per	Installation Groups			
				1	2	3	4
MATERIAL							
Box and support	1		E				
#12-3 Type AC Cable	20 - 30 ft.		C Ft.				
Two-circuit receptacle and plate	1		E				
Miscellaneous	Lot						
TOTAL MATERIAL COST							
TOTAL LABOR COST - GROUP 1	0.75 WH		Hr				
GROUP 2	1.10 WH		Hr				
GROUP 3	1.45 WH		Hr				
GROUP 4	1.80 WH		Hr				
DIRECT JOB EXPENSE							
TOTAL PRIME COST				$	$	$	$

Duplex Receptacle – Weatherproof

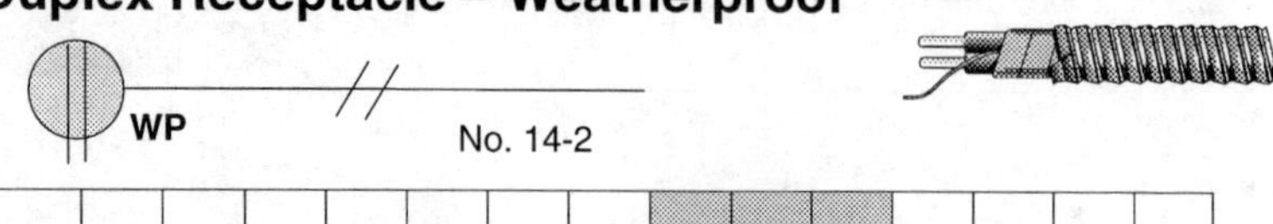

Cost Item	Quantity	Price or Rate	Per	Installation Groups 1	2	3	4
MATERIAL							
Box and support	1		E				
#14-2 Type AC Cable	20 - 30 ft.		C Ft.				
Receptacle and weatherproof plate	1		E				
Miscellaneous	Lot						
TOTAL MATERIAL COST							
TOTAL LABOR COST - GROUP 1	0.75 WH		Hr				
GROUP 2	1.10 WH		Hr				
GROUP 3	1.55 WH		Hr				
GROUP 4	1.95 WH		Hr				
DIRECT JOB EXPENSE							
TOTAL PRIME COST				$	$	$	$

Duplex Receptacle – Weatherproof

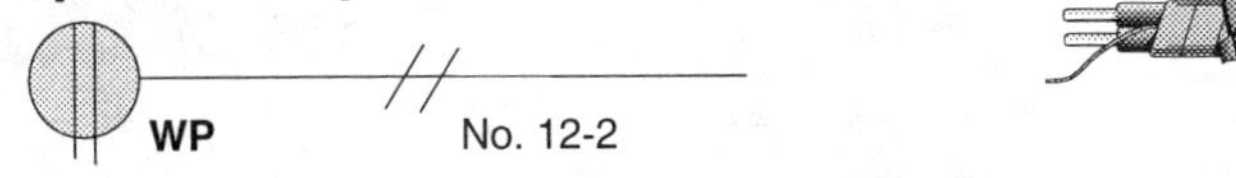

Cost Item	Quantity	Price or Rate	Per	Installation Groups 1	2	3	4
MATERIAL							
Box and support	1		E				
#12-2 Type AC Cable	20 - 30 ft.		C Ft.				
Receptacle and weatherproof plate	1		E				
Miscellaneous	Lot						
TOTAL MATERIAL COST							
TOTAL LABOR COST - GROUP 1	0.85 WH		Hr				
GROUP 2	1.20 WH		Hr				
GROUP 3	1.65 WH		Hr				
GROUP 4	2.05 WH		Hr				
DIRECT JOB EXPENSE							
TOTAL PRIME COST				$	$	$	$

Duplex Receptacle – GFCI

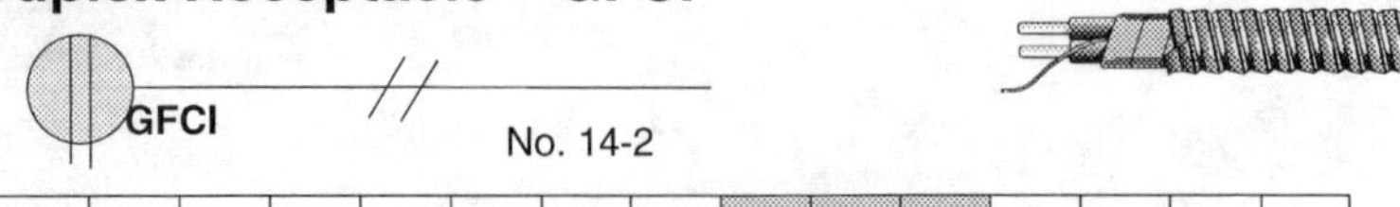

No. 14-2

Cost Item	Quantity	Price or Rate	Per	Installation Groups 1	2	3	4
MATERIAL							
Box and support	1		E				
#14-2 Type AC Cable	20 - 30 ft.		C Ft.				
GFCI receptacle and plate or	1		E				
GFCI circuit breaker and regular recept.	1		E				
Miscellaneous	Lot						
TOTAL MATERIAL COST							
TOTAL LABOR COST - GROUP 1	0.75 WH		Hr				
GROUP 2	1.05 WH		Hr				
GROUP 3	1.35 WH		Hr				
GROUP 4	1.65 WH		Hr				
DIRECT JOB EXPENSE							
TOTAL PRIME COST				$	$	$	$

Duplex Receptacle – GFCI

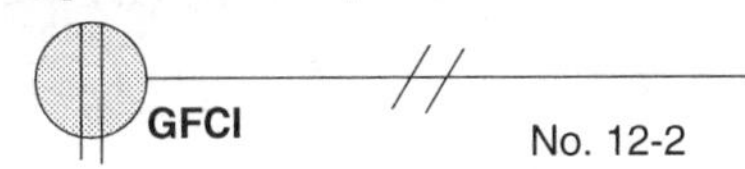

Cost Item	Quantity	Price or Rate	Per	Installation Groups 1	2	3	4
MATERIAL							
Box and support	1		E				
#12-2 Type AC Cable	20 - 30 ft.		C Ft.				
GFCI receptacle and plate or	1		E				
GFCI circuit breaker and regular recept.	1		E				
Miscellaneous	Lot						
TOTAL MATERIAL COST							
TOTAL LABOR COST - GROUP 1	0.80 WH		Hr				
GROUP 2	1.15 WH		Hr				
GROUP 3	1.40 WH		Hr				
GROUP 4	1.70 WH		Hr				
DIRECT JOB EXPENSE							
TOTAL PRIME COST				$	$	$	$

Duplex Receptacle – Floor Mounted

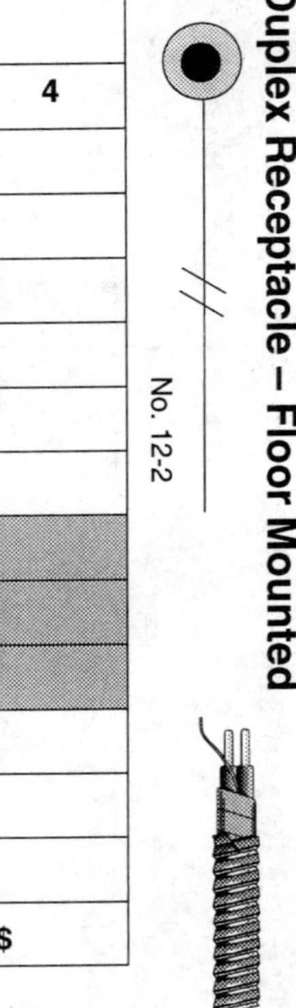

No. 12-2

Cost Item	Quantity	Price or Rate	Per	Installation Groups			
				1	2	3	4
MATERIAL							
Box and support	1		E				
#12-2 Type AC Cable	20 - 30 ft.		C Ft.				
Floor receptacle and cover	1		E				
Miscellaneous	Lot						
TOTAL MATERIAL COST							
TOTAL LABOR COST - GROUP 1	1.25 WH		Hr				
GROUP 2	1.85 WH		Hr				
GROUP 3	2.45 WH		Hr				
GROUP 4	3.10 WH		Hr				
DIRECT JOB EXPENSE							
TOTAL PRIME COST				$	$	$	$

Duplex Receptacle – Clock Hanger

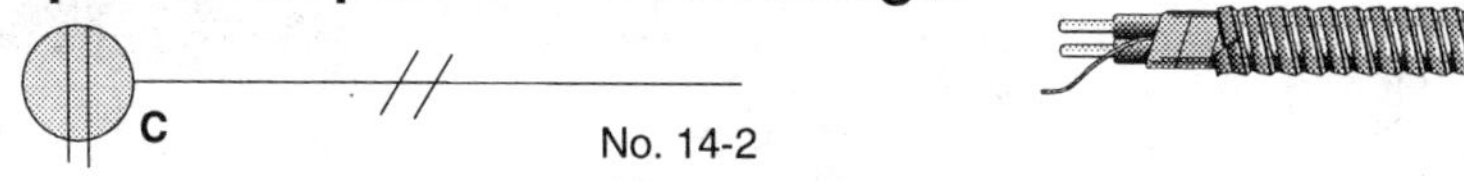

Cost Item	Quantity	Price or Rate	Per	Installation Groups 1	2	3	4
MATERIAL							
Box and support	1		E				
#14-2 Type AC Cable	20 - 30 ft.		C Ft.				
Clock hanger receptacle and cover	1		E				
Miscellaneous	Lot						
TOTAL MATERIAL COST							
TOTAL LABOR COST - GROUP 1	0.65 WH		Hr				
GROUP 2	0.95 WH		Hr				
GROUP 3	1.35 WH		Hr				
GROUP 4	1.70 WH		Hr				
DIRECT JOB EXPENSE							
TOTAL PRIME COST				$	$	$	$

Ceiling Fan Receptacle

No. 12-2

Cost Item	Quantity	Price or Rate	Per	Installation Groups			
				1	2	3	4
MATERIAL							
Ceiling fan box and support	1		E				
#12-2 Type AC Cable	20 - 30 ft.		C Ft.				
Ceiling fan receptacle and cover	1		E				
Miscellaneous	Lot						
TOTAL MATERIAL COST							
TOTAL LABOR COST - GROUP 1	1.40 WH		Hr				
GROUP 2	2.00 WH		Hr				
GROUP 3	2.60 WH		Hr				
GROUP 4	3.20 WH		Hr				
DIRECT JOB EXPENSE							
TOTAL PRIME COST				$	$	$	$

Wall Switch – Single-Pole

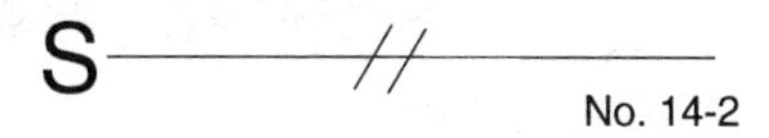

Cost Item	Quantity	Price or Rate	Per	Installation Groups 1	2	3	4
MATERIAL							
Box and support	1		E				
#14-2 Type AC Cable	20 - 30 ft.		C Ft.				
Single-pole switch and plate	1		E				
Miscellaneous	Lot						
TOTAL MATERIAL COST							
TOTAL LABOR COST - GROUP 1	0.55 WH		Hr				
GROUP 2	0.85 WH		Hr				
GROUP 3	1.15 WH		Hr				
GROUP 4	1.45 WH		Hr				
DIRECT JOB EXPENSE							
TOTAL PRIME COST				$	$	$	$

Wall Switch – Single-Pole

S ——//——

No. 12-2

Cost Item	Quantity	Price or Rate	Per	Installation Groups 1	2	3	4
MATERIAL							
Box and support	1		E				
#12-2 Type AC Cable	20 - 30 ft.		C Ft.				
Single-pole switch with plate	1		E				
Miscellaneous	Lot						
TOTAL MATERIAL COST							
TOTAL LABOR COST - GROUP 1	0.65 WH		Hr				
GROUP 2	0.95 WH		Hr				
GROUP 3	1.25 WH		Hr				
GROUP 4	1.55 WH		Hr				
DIRECT JOB EXPENSE							
TOTAL PRIME COST				$	$	$	$

Wall Switch with Pilot Light

Cost Item	Quantity	Price or Rate	Per	Installation Groups 1	2	3	4
MATERIAL							
Box and support	1		E				
#14-3 Type AC Cable	20 - 30 ft.		C Ft.				
Combination switch/pilot light with plate	1		E				
Miscellaneous	Lot						
TOTAL MATERIAL COST							
TOTAL LABOR COST - GROUP 1	0.65 WH		Hr				
GROUP 2	1.00 WH		Hr				
GROUP 3	1.35 WH		Hr				
GROUP 4	1.70 WH		Hr				
DIRECT JOB EXPENSE							
TOTAL PRIME COST				$	$	$	$

Wall Switch with Pilot Light

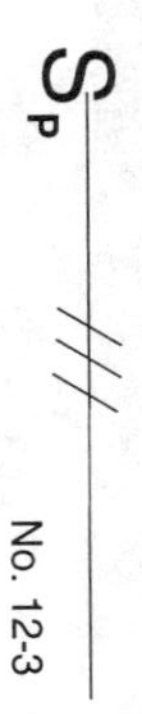

No. 12-3

Cost Item	Quantity	Price or Rate	Per	Installation Groups 1	2	3	4
MATERIAL							
Box and support	1		E				
#12-3 Type AC Cable	20 - 30 ft.		C Ft.				
Combination switch/pilot light with plate	1		E				
Miscellaneous	Lot						
TOTAL MATERIAL COST							
TOTAL LABOR COST - GROUP 1	0.75 WH		Hr				
GROUP 2	1.10 WH		Hr				
GROUP 3	1.45 WH		Hr				
GROUP 4	1.80 WH		Hr				
DIRECT JOB EXPENSE							
TOTAL PRIME COST				$	$	$	$

Wall Switch – Three-Way

Cost Item	Quantity	Price or Rate	Per	Installation Groups 1	2	3	4
MATERIAL							
Box and support	1		E				
#14-3 Type AC Cable	25 - 35 ft.		C Ft.				
Three-way switch with plate	1		E				
Miscellaneous	Lot						
TOTAL MATERIAL COST							
TOTAL LABOR COST - GROUP 1	0.75 WH		Hr				
GROUP 2	1.15 WH		Hr				
GROUP 3	1.55 WH		Hr				
GROUP 4	1.95 WH		Hr				
DIRECT JOB EXPENSE							
TOTAL PRIME COST				$	$	$	$

Wall Switch – Three-Way

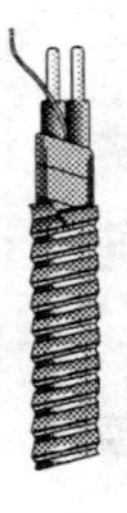

No. 12-3

Cost Item	Quantity	Price or Rate	Per	Installation Groups			
				1	2	3	4
MATERIAL							
Box and support	1		E				
#12-3 Type AC Cable	25 - 35 ft.		C Ft.				
Three-way switch with plate	1		E				
Miscellaneous	Lot						
TOTAL MATERIAL COST							
TOTAL LABOR COST - GROUP 1	0.85 WH		Hr				
GROUP 2	1.25 WH		Hr				
GROUP 3	1.65 WH		Hr				
GROUP 4	2.10 WH		Hr				
DIRECT JOB EXPENSE							
TOTAL PRIME COST				$	$	$	$

Wall Switch – Four-Way

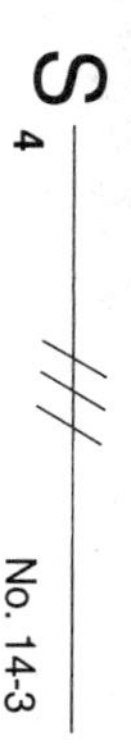

Cost Item	Quantity	Price or Rate	Per	Installation Groups			
				1	2	3	4
MATERIAL							
Box and support	1		E				
#14-3 Type AC Cable	35 - 50 ft.		C Ft.				
Four-way switch and plate	1		E				
Miscellaneous	Lot						
TOTAL MATERIAL COST							
TOTAL LABOR COST - GROUP 1	0.75 WH		Hr				
GROUP 2	1.15 WH		Hr				
GROUP 3	1.55 WH		Hr				
GROUP 4	1.95 WH		Hr				
DIRECT JOB EXPENSE							
TOTAL PRIME COST				$	$	$	$

Wall Switch – Four-Way

S_4 ///

No. 12-3

Cost Item	Quantity	Price or Rate	Per	Installation Groups 1	2	3	4
MATERIAL							
Box and support	1		E				
#12-3 Type AC Cable	35 - 50 ft.		C Ft.				
Four-way switch with plate	1		E				
Miscellaneous	Lot						
TOTAL MATERIAL COST							
TOTAL LABOR COST - GROUP 1	0.85 WH		Hr				
GROUP 2	1.25 WH		Hr				
GROUP 3	1.65 WH		Hr				
GROUP 4	2.10 WH		Hr				
DIRECT JOB EXPENSE							
TOTAL PRIME COST				$	$	$	$

Dimmer Switch/Control

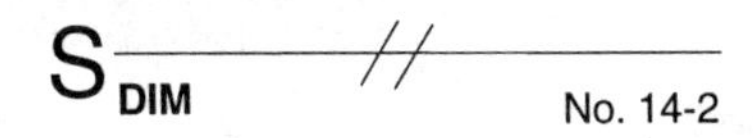

Cost Item	Quantity	Price or Rate	Per	Installation Groups 1	2	3	4
MATERIAL							
Box and support	1		E				
#14-2 Type AC Cable	20 - 35 ft.		C Ft.				
Dimmer control and cover	1		E				
Miscellaneous	Lot						
TOTAL MATERIAL COST							
TOTAL LABOR COST - GROUP 1	0.95 WH		Hr				
GROUP 2	1.45 WH		Hr				
GROUP 3	1.95 WH		Hr				
GROUP 4	2.40 WH		Hr				
DIRECT JOB EXPENSE							
TOTAL PRIME COST				$	$	$	$

Dimmer Switch/Control

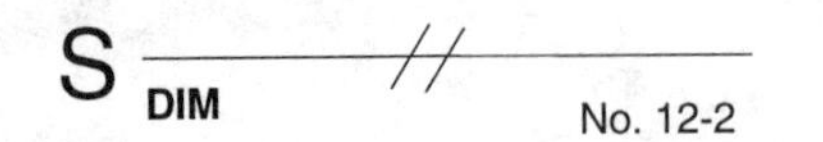

No. 12-2

Cost Item	Quantity	Price or Rate	Per	Installation Groups 1	2	3	4
MATERIAL							
Box and support	1		E				
#12-2 Type AC Cable	15 - 35 ft.		C Ft.				
Dimmer control with cover	1		E				
Miscellaneous	Lot						
TOTAL MATERIAL COST							
TOTAL LABOR COST - GROUP 1	1.05 WH		Hr				
GROUP 2	1.55 WH		Hr				
GROUP 3	2.05 WH		Hr				
GROUP 4	2.55 WH		Hr				
DIRECT JOB EXPENSE							
TOTAL PRIME COST				$	$	$	$

Selling-Price Tables

The selling-price tables to follow are designed for quick unit pricing of the various outlets found in this chapter. Before arriving at a selling price, the preceding tables must be completed (filled in with appropriate prices). Then the prime cost determined for each wiring situation in the preceding tables should be entered in the corresponding selling-price tables to follow. Once the prime cost has been entered in each selling-price table, overhead and profit factors are calculated to arrive at a total unit selling price as described in detail in Chapter 2.

Surface-Mounted Lighting Outlet	14-2	GROUPS			
		1	2	3	4
Total Prime Cost		$	$	$	$
Overhead	_% of prime cost				
Total Cost		$	$	$	$
Profit	_% of total cost				
Selling Price		$	$	$	$

Surface-Mounted Lighting Outlet	12-2	GROUPS			
		1	2	3	4
Total Prime Cost		$	$	$	$
Overhead	_% of prime cost				
Total Cost		$	$	$	$
Profit	_% of total cost				
Selling Price		$	$	$	$

Recessed Lighting Outlet	14-2	GROUPS			
		1	2	3	4
Total Prime Cost		$	$	$	$
Overhead	_% of prime cost				
Total Cost		$	$	$	$
Profit	_% of total cost				
Selling Price		$	$	$	$

Recessed Lighting Outlet	12-2	GROUPS			
		1	2	3	4
Total Prime Cost		$	$	$	$
Overhead	_% of prime cost				
Total Cost		$	$	$	$
Profit	_% of total cost				
Selling Price		$	$	$	$

Wall-Mounted Lighting Outlet	14-2	GROUPS			
		1	2	3	4
Total Prime Cost		$	$	$	$
Overhead	_% of prime cost				
Total Cost		$	$	$	$
Profit	_% of total cost				
Selling Price		$	$	$	$

Wall-Mounted Lighting Outlet	12-2	GROUPS			
		1	2	3	4
Total Prime Cost		$	$	$	$
Overhead	_% of prime cost				
Total Cost		$	$	$	$
Profit	_% of total cost				
Selling Price		$	$	$	$

Duplex Receptacle 2-Wire Circuit	14-2	GROUPS			
		1	2	3	4
Total Prime Cost		$	$	$	$
Overhead	_% of prime cost				
Total Cost		$	$	$	$
Profit	_% of total cost				
Selling Price		$	$	$	$

Duplex Receptacle 2-Wire Circuit	12-2	GROUPS			
		1	2	3	4
Total Prime Cost		$	$	$	$
Overhead	_% of prime cost				
Total Cost		$	$	$	$
Profit	_% of total cost				
Selling Price		$	$	$	$

Duplex Receptacle 3-Wire Circuit	14-3	GROUPS			
		1	2	3	4
Total Prime Cost		$	$	$	$
Overhead	_% of prime cost				
Total Cost		$	$	$	$
Profit	_% of total cost				
Selling Price		$	$	$	$

Duplex Receptacle 3-Wire Circuit	12-3	**GROUPS**			
		1	**2**	**3**	**4**
Total Prime Cost		$	$	$	$
Overhead	_% of prime cost				
Total Cost		$	$	$	$
Profit	_% of total cost				
Selling Price		$	$	$	$

Duplex Receptacle Split-Wired	14-3	**GROUPS**			
		1	**2**	**3**	**4**
Total Prime Cost		$	$	$	$
Overhead	_% of prime cost				
Total Cost		$	$	$	$
Profit	_% of total cost				
Selling Price		$	$	$	$

Duplex Receptacle Split-Wired	12-3	**GROUPS**			
		1	**2**	**3**	**4**
Total Prime Cost		$	$	$	$
Overhead	_% of prime cost				
Total Cost		$	$	$	$
Profit	_% of total cost				
Selling Price		$	$	$	$

Duplex Receptacle Weatherproof	14-2	GROUPS			
		1	2	3	4
Total Prime Cost		$	$	$	$
Overhead	_% of prime cost				
Total Cost		$	$	$	$
Profit	_% of total cost				
Selling Price		$	$	$	$

Duplex Receptacle Weatherproof	12-2	GROUPS			
		1	2	3	4
Total Prime Cost		$	$	$	$
Overhead	_% of prime cost				
Total Cost		$	$	$	$
Profit	_% of total cost				
Selling Price		$	$	$	$

Duplex Receptacle GFCI	14-2	GROUPS			
		1	2	3	4
Total Prime Cost		$	$	$	$
Overhead	_% of prime cost				
Total Cost		$	$	$	$
Profit	_% of total cost				
Selling Price		$	$	$	$

Duplex Receptacle GFCI	12-2	GROUPS			
		1	2	3	4
Total Prime Cost		$	$	$	$
Overhead	_% of prime cost				
Total Cost		$	$	$	$
Profit	_% of total cost				
Selling Price		$	$	$	$

Duplex Receptacle Floor Mounted	12-2	GROUPS			
		1	2	3	4
Total Prime Cost		$	$	$	$
Overhead	_% of prime cost				
Total Cost		$	$	$	$
Profit	_% of total cost				
Selling Price		$	$	$	$

Duplex Receptacle Clock Hanger	14-2	GROUPS			
		1	2	3	4
Total Prime Cost		$	$	$	$
Overhead	_% of prime cost				
Total Cost		$	$	$	$
Profit	_% of total cost				
Selling Price		$	$	$	$

Ceiling Fan Receptacle	12-2	GROUPS 1	2	3	4
Total Prime Cost		$	$	$	$
Overhead	_% of prime cost				
Total Cost		$	$	$	$
Profit	_% of total cost				
Selling Price		$	$	$	$

Wall Switch Single-Pole	14-2	GROUPS 1	2	3	4
Total Prime Cost		$	$	$	$
Overhead	_% of prime cost				
Total Cost		$	$	$	$
Profit	_% of total cost				
Selling Price		$	$	$	$

Wall Switch Single-Pole	12-2	GROUPS 1	2	3	4
Total Prime Cost		$	$	$	$
Overhead	_% of prime cost				
Total Cost		$	$	$	$
Profit	_% of total cost				
Selling Price		$	$	$	$

Wall Switch With Pilot Light	14-3	GROUPS			
		1	2	3	4
Total Prime Cost		$	$	$	$
Overhead	_% of prime cost				
Total Cost		$	$	$	$
Profit	_% of total cost				
Selling Price		$	$	$	$

Wall Switch With Pilot Light	12-3	GROUPS			
		1	2	3	4
Total Prime Cost		$	$	$	$
Overhead	_% of prime cost				
Total Cost		$	$	$	$
Profit	_% of total cost				
Selling Price		$	$	$	$

Wall Switch Three-Way	14-3	GROUPS			
		1	2	3	4
Total Prime Cost		$	$	$	$
Overhead	_% of prime cost				
Total Cost		$	$	$	$
Profit	_% of total cost				
Selling Price		$	$	$	$

Wall Switch Three-Way	12-3	GROUPS			
		1	2	3	4
Total Prime Cost		$	$	$	$
Overhead	_% of prime cost				
Total Cost		$	$	$	$
Profit	_% of total cost				
Selling Price		$	$	$	$

Wall Switch Four-Way	14-3	GROUPS			
		1	2	3	4
Total Prime Cost		$	$	$	$
Overhead	_% of prime cost				
Total Cost		$	$	$	$
Profit	_% of total cost				
Selling Price		$	$	$	$

Wall Switch Four-Way	12-3	GROUPS			
		1	2	3	4
Total Prime Cost		$	$	$	$
Overhead	_% of prime cost				
Total Cost		$	$	$	$
Profit	_% of total cost				
Selling Price		$	$	$	$

Dimmer Switch/Control	14-2	GROUPS			
		1	2	3	4
Total Prime Cost		$	$	$	$
Overhead	_% of prime cost				
Total Cost		$	$	$	$
Profit	_% of total cost				
Selling Price		$	$	$	$

Dimmer Switch/Control	12-2	GROUPS			
		1	2	3	4
Total Prime Cost		$	$	$	$
Overhead	_% of prime cost				
Total Cost		$	$	$	$
Profit	_% of total cost				
Selling Price		$	$	$	$

Index

Page numbers in italics indicate selling-price table locations

S

T

U-V

W

Other Practical References

Audiotapes: Estimating Electrical Work

Listen to Trade Service's two-day seminar and study electrical estimating at your own speed for a fraction of the cost of attending the actual seminar. You'll learn what to expect from specifications, how to adjust labor units from a price book to your job, how to make an accurate take-off from the plans, and how to spot hidden costs that other estimators may miss. *Includes six 30-minute tapes, a workbook that includes price sheets, specification sheet, bid summary, estimate recap sheet, blueprints used in the actual seminar, and blank forms for your own use.* **$65.00**

Electrician's Exam Preparation Guide

Need help in passing the apprentice, journeyman, or master electrician's exam? This is a book of questions and answers based on actual electrician's exams over the last few years. Almost a thousand multiple-choice questions — exactly the type you'll find on the exam — cover every area of electrical installation: electrical drawings, services and systems, transformers, capacitors, distribution equipment, branch circuits, feeders, calculations, measuring and testing, and more. It gives you the correct answer, an explanation, and where to find it in the code. Also tells how to apply for the test, how best to study, and what to expect on examination day. **320 pages, 8½ x 11, $23.00**

Estimating Electrical Construction

Like taking a class in how to estimate materials and labor for residential and commercial electrical construction. Written by an A.S.P.E. National Estimator of the Year, it teaches you how to use labor units, the plan take-off, and the bid summary to make an accurate estimate, how to deal with suppliers, use pricing sheets, and modify labor units. Provides extensive labor unit tables and blank forms for your next electrical job.
272 pages, 8½ x 11, $19.00

Electrical Blueprint Reading Revised

Shows how to read and interpret electrical drawings, wiring diagrams, and specifications for constructing electrical systems. Shows how a typical lighting and power layout would appear on a plan, and explains what to do to execute the plan. Describes how to use a panelboard or heating schedule, and includes typical electrical specifications.
208 pages, 8½ x 11, $18.00

National Electrical Estimator

This year's prices for installation of all common electrical work: conduit, wire, boxes, fixtures, switches, outlets, loadcenters, panelboards, raceway, duct, signal systems, and more. Provides material costs, manhours per unit, and total installed cost. Explains what you should know to estimate each part of an electrical system. Includes an electronic version of the book on computer disk with a stand-alone *Windows* estimating program **FREE** on a 3½" high-density (1.44 Mb) disk. If you need 5¼" high-density or 3½ double- density disks add $10 extra. **512 pages, 8½ x 11, $31.75. Revised annually**

Illustrated Guide to the 1993 National Electrical Code®

This fully-illustrated guide offers a quick and easy visual reference for installing electrical systems. Whether you're installing a new system or repairing an old one, you'll appreciate the simple explanations written by a code expert, and the detailed, intricately-drawn and labeled diagrams. A real time-saver when it comes to deciphering the current *NEC*. **256 pages, 8½ x 11, $26.75**

Commercial Electrical Wiring

Make the transition from residential to commercial electrical work. Here are wiring methods, spec reading tips, load calculations and everything you need for making the transition to commercial work: commercial construction documents, load calculations, electric services, transformers, overcurrent protection, wiring methods, raceway, boxes and fittings, wiring devices, conductors, electric motors, relays and motor controllers, special occupancies, and safety requirements. This book is written to help any electrician break into the lucrative field of commercial electrical work. **320 pages, 8½ x 11, $27.50**

Residential Electrical Design Revised

If you've ever had to draw up an electrical plan for an addition, or add corrections to an existing plan, you know how complicated it can get. And how many electrical plans — no matter how well designed — fit the reality of what the homeowner wants? Here you'll find everything you need to know about blueprints, what the NEC requires, how to size electric service, calculate and size loads and conductors, install ground-fault circuit interrupters, ground service entrances, and recommended wiring methods. It covers branch circuit layout, how to analyze existing lighting layouts and install outdoor lighting, methods for remote-control switching, residential HVAC systems and controls, and more. **256 pages, 8½ x 11, $22.50**

NO POSTAGE
NECESSARY
IF MAILED
IN THE
UNITED STATES

BUSINESS REPLY MAIL
FIRST CLASS MAIL PERMIT NO. 271 CARLSBAD, CA

POSTAGE WILL BE PAID BY ADDRESSEE

Craftsman Book Company
6058 Corte del Cedro
P.O. Box 6500
Carlsbad, CA 92018-9974

NO POSTAGE
NECESSARY
IF MAILED
IN THE
UNITED STATES

BUSINESS REPLY MAIL
FIRST CLASS MAIL PERMIT NO. 271 CARLSBAD, CA

POSTAGE WILL BE PAID BY ADDRESSEE

Craftsman Book Company
6058 Corte del Cedro
P.O. Box 6500
Carlsbad, CA 92018-9974